Heterogeneous Electrode
Processes and
Localized Corrosion

WILEY SERIES IN CORROSION

R. Winston Revie, Series Editor

Heterogeneous Electrode Processes and Localized Corrosion

YONGJUN TAN

A JOHN WILEY & SONS, INC., PUBLICATION

For general information on our other products and services or for technical support, please contact our Customer Care Department within the United States at (800) 762-2974, outside the United States at (317) 572-3993 or fax (317) 572-4002.

Wiley also publishes its books in a variety of electronic formats. Some content that appears in print may not be available in electronic formats. For more information about Wiley products, visit our web site at www.wiley.com.

Library of Congress Cataloging-in-Publication Data:

Tan, Yongjun, 1963–
   Heterogeneous electrode processes and localized corrosion / Yongjun Tan.
      pages cm. – (Wiley series in corrosion)
   Includes bibliographical references and index.
   ISBN 978-0-470-64795-0 (hardback)
   1. Corrosion and anti-corrosives. 2. Heterogeneous catalysis. 3. Electrocatalysis. I. Title.
   TA418.74.T36 2013
   620.1′1223–dc23

                                                2012020902

Printed in the United States of America

10 9 8 7 6 5 4 3 2 1

Contents

v

Preface

Electrochemistry is an enabling science that since Galvani's 1780s experiment on frog legs and Volta's nineteenth-century invention of the battery has facilitated a wide spectrum of scientific discoveries and industrial processes. These include electrolysis, electroplating, metal winning, hydrometallurgy, corrosion prevention, passivation, batteries, fuel cells, solar cells, environmental and biological sensors, organic and inorganic electrosynthesis, electrophoresis, effluent remediation, and neurochemistry. Electrochemical theories and models have also been developed, including the most fundamental relationships, such as Faraday's law, the Nernst equation, and the Bulter–Volmer formulation. However, traditional electrochemistry has a major weakness—there is a "missing link" between its ideal uniform electrode model and practical heterogeneous electrode processes because of the phenomenon of electrochemical heterogeneity.

Electrochemical heterogeneity refers to the localization of chemical and electrochemical reactions over an electrode surface. In electrochemical reactions the reactants do not need to be near each other spatially as normal chemical reactions do; they collide separately with spatially separated electronic conductors. This characteristic permits distinct separation of electrode reactions over an electrode surface, leading to the localization of electrode chemistry, reaction thermodynamics, and kinetics. Most frequently, Electrochemical heterogeneity is initiated from preexisting electrode inhomogeneity (i.e., spatial, physical, compositional, metallurgical, or chemical nonuniformities existing on electrode surfaces). It could also be initiated from an ideally homogeneous surface, especially if the surface were exposed to a nonuniform environment. Electrochemical heterogeneity is a ubiquitous phenomenon that plays critical roles in many difficult but important issues in electrochemical science and engineering, such as localized corrosion, porous electrodeposition, uneven electrodissolution, electrochemical noise, and various forms of dispersion in voltammetry. Electrochemical heterogeneity has not been confronted effectively in traditional electrochemical science research and has not been covered sufficiently in the electrochemical and corrosion science literature. It is evident that classical theories describing the Faradaic electron transfer processes in dynamic electrochemistry and traditional methods of electrochemical measurement are based on the assumption that the electrode surface is homogeneous. Initial experimental studies used to validate traditional electrochemical theories and methods commonly employed mercury drop electrodes, where a truly homogeneous surface is indeed likely to be achieved. However, solid electrodes that are in practical use today generally have inhomogeneous surfaces, where electrochemical heterogeneity could initiate, propagate, or terminate dynamically. Indeed, there is a major theoretical and technological gap between conventional electrochemistry over uniform surfaces and heterogeneous electrochemistry over inhomogeneous surfaces.

I was first attracted to the issue of electrochemical heterogeneity in 1988 when I was mystified by difficulties in repeating my impedance spectra of coated electrodes. Over the past two decades I have been fortunate to be able to continue my thinking and research on this issue, leading to the development and application of an electrochemically integrated multielectrode array: the wire beam electrode. My colleagues and I have carried out numerous experiments to study a variety of heterogeneous electrode processes. Our experimental findings indicate that electrochemists may need to recognize more widely the fundamental significance of electrode inhomogeneity and electrochemical heterogeneity.

This book is probably the first to focus on electrode inhomogeneity, electrochemical heterogeneity, and their effects on nonuniform electrode processes. Attempts are made to critically review various forms of techniques that have been applied for probing localized electrode–electrolyte interfaces, in particular scanning probe techniques such as the scanning Kelvin probe, the scanning Kelvin probe force microscopy, the scanning vibrating electrode technique, local electrochemical impedance spectroscopy, and scanning electrochemical microscopy. Special attention is focused on localized corrosion experiments designed using the wire beam electrode as a key tool. Case studies presented in this book to illustrate innovative experiments are based primarily on published data with which I have substantial firsthand experience. I acknowledge work by several other research groups that have reported innovative and interesting experiments using various forms of electrode arrays.

I express my sincere thanks to colleagues and co-workers who have collaborated with me over the past two decades, especially (in chronological order) Cuilan Wu, Xuejun Zhou, Shiti Yu, Stuart Bailey, Yadran Marinovich, Brian Kinsella, Alex Lowe, Tie Liu, Naing Naing Aung, Ting Wang, Kim Yong Lim, Bernice Zee,

Alan Bond, Chong Yong Lee, Mauro Mocerino, Tristan Paterson, Young Fwu, Kriti Bhardwaj, Bruce Hinton, and Maria Forsyth. I thank them and many other colleagues who have taught me patiently over the years. I also thank my family for giving me encouragement, time, and support throughout the years.

YOUNGJUN TAN

1

Homogeneous Electrode Models and Uniform Corrosion Measurements

Electrochemical reaction is a special form of chemical reaction that occurs in reaction devices: that is, electrochemical cells. Since Galvanis' 1780's experiment on frog legs and Volta's nineteenth-century invention of the voltaic pile, numerous types of electrochemical cells, such as batteries, fuel cells, solar cells, galvanic cells, and electroplating and electrowinning baths, have been invented by electrochemists, facilitating a wide spectrum of industrial processes and scientific discoveries. Many electrochemistry-based technologies, including electrolysis, electrosynthesis, electrocatalysis; electroplating, metal winning, hydrometallurgy, electrophoresis, effluent remediation, electrochemical energy conversion and storage, cathodic and anodic protection, and environmental, biological, and corrosion sensors, have been developed for various industrial applications. Electrochemistry also provides a theoretical basis for understanding and explaining natural phenomena in scientific fields such as corrosion science and neurochemistry.

Fundamentally, electrochemical cells are based on two basic electrochemical devices: the galvanic cell (Figure 1.1a) and the electrolytic cell (Figure 1.1b). All electrochemical cells consist of an anode, a cathode, an electrically conductive path, and an ionically conductive path. Electrodes are basic components that provide the interface between an electronic conductor, usually a metal, and an ionic conductor, usually an electrolytic solution.

Electrochemical reactions in a galvanic or electrolytic cell always involve two half-cell reactions occurring simultaneously over separated anode and cathode

---

Heterogeneous Electrode Processes and Localized Corrosion, First Edition. Yongjun Tan.
© 2013 John Wiley & Sons, Inc. Published 2013 by John Wiley & Sons, Inc.

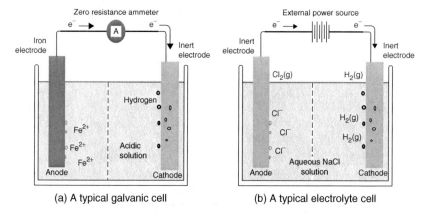

(a) A typical galvanic cell   (b) A typical electrolyte cell

Figure 1.1 Basic electrochemical devices.

surfaces. Anodic oxidation reactions occur at the anode–solution interface across which species lose electrons, while cathodic reduction reactions occur at the cathode–solution interface across which species gain electrons. The mass transfer, chemical reaction, and physical state change are the sequential steps of electrochemical processes occurring at the anodic and cathodic half-cells. In a *galvanic cell* (Figure 1.1a), electrochemical reactions occur spontaneously at the anodes and cathodes when they are connected externally by a conductor. Galvanic cells are often employed as batteries and fuel cells to convert chemical energy into electrical energy. A galvanic cell also provides a starting point for the development of metal corrosion theories and corrosion prevention technologies. In an *electrolytic cell* (Figure 1.1b), electrochemical reactions are driven by the imposition of an external voltage greater than the open-circuit potential of the cell. Electrolytic cells are widely employed in such industrial processes as electrolysis, metal winning, electroplating, and cathodic protection.

An electrochemical reaction is a chemical reaction in nature, and therefore all factors affecting normal chemical reactions, such as the chemical nature of reactants, the concentrations of the reactants, the temperature, the ability of reactants to come into contact with each other, and the availability of rate-accelerating or rate-decelerating agents, affect electrochemical reactions. However, it should be pointed out that there is a fundamental difference between normal chemical reactions and electrochemical reactions. An electrochemical reaction occurs at the electrode–solution interface, often driven by voltage applied externally. In an electrochemical reaction the reactants do not need to be near each other spatially as normal chemical reactions do; they collide individually with spatially separated anodes and cathodes. For this reason, electrochemical reactions are affected significantly by the structure and property of electrode–solution interfaces, movements of ions and electrons between electrodes, and the voltage applied externally.

In this chapter we provide an overview of models illustrating homogeneous electrode–solution interfacial structure and theories describing electrode processes with and without the effects of voltage applied externally. Particular focus is on a mixed electrode model that explains uniform corrosion. Electrochemical and corrosion principles are reviewed; however, a detailed description of traditional electrochemical and corrosion theories is not attempted because these have already been discussed by many authors, among them Bard and Faulkner [1], Bockris et al. [2], Fontana [3], Mansfeld [4], and Evans [5]. Attempts are made to discuss limitations in traditional electrochemical methods and factors that may cause a change from a uniform corrosion mechanism to a localized form, leading to the concepts of electrode inhomogeneity and electrochemical heterogeneity.

Carbon dioxide (CO_2) corrosion of iron in sodium chloride (NaCl)-containing solutions under ambient temperature and atmospheric pressure conditions is used as a case to illustrate typical uniform corrosion processes occurring in practical industrial systems. Electrochemical measurements are described through their applications in investigating the rates and interfacial structures of steel electrodes in CO_2 corrosion environments, with and without the present of corrosion inhibitors.

1.1 HOMOGENEOUS ELECTRODES AND TRADITIONAL ELECTROCHEMICAL METHODS

Electrochemical reactions occur at the interface of an electrode and a solution. Knowledge of the structure and property of an electrode–solution interface is critical for understanding electrochemical reactions and their thermodynamics and kinetics. It is well known that the electrode–solution interface is affected by electrode surface roughness and scratches, impurities, mill scales, surface flaws, metallurgical defects, precipitated phases, grain boundaries, dislocation arrays, localized stresses, selective dissolution, and damage to passive films. It is also understood that an electrode–solution interface is affected by factors such as surface chemical adsorption and desorption, mass transfer, and liquid flow. However, unfortunately, direct observation of an electrode–solution interface is very difficult, if not impossible. Instead, theoretical models are employed to help in addressing issues relating to the structure, properties, and reactivity of an electrode–solution interfacial region and its components.

Figure 1.2 is a simplified model of a conventional uniform electrode–solution interface. In this model it is assumed that an electrochemical double layer exists in the vicinity of an electrode surface, consisting of inner and outer Helmholtz layers. It is also assumed that the electrode surface is chemically and electrochemically homogeneous and that electrode reactions are distributed uniformly over the electrode surface.

This homogeneous electrode–solution interface model significantly simplifies the analysis of electrochemical processes. Under this homogeneous electrode assumption, electrochemical properties at any location of an electrode surface can be considered to be identical to that of the entire electrode surface. For example, it allows the application of Faraday's law of electrolysis [1] to determine the thickness of electrodeposits because electrodeposition current can be considered

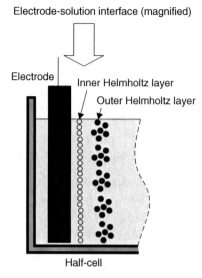

Figure 1.2 Simplified electrode–solution interface model.

as being uniformly distributed over an electrode surface. This assumption is also employed as a prerequisite for fundamental electrochemical thermodynamic and kinetic relationships, including the Nernst equation and the Bulter–Volmer formulation and also for traditional electrochemical methods such as electrode potential and polarization measurements [1–5].

This electrode–solution interface model is applicable to both the anode and cathode of a galvanic cell or an electrolytic cell. In a galvanic cell, such as the iron corrosion cell shown in Figure 1.1a, actual reactions occurring at its anodic and cathodic interfaces are determined by the electrochemical thermodynamics of chemical species in the cell. Electrode potential is a thermodynamic parameter that reveals the tendency of a chemical species to be reduced in an electrochemical reaction. Figure 1.3 illustrates a traditional setup for determining the standard reduction potential of an iron electrode in 1.00 M Fe^{2+} solution using a standard hydrogen reference electrode and a high-resistance voltmeter. This laboratory method determines the reduction potential of 1.00 M Fe^{2+} against the reduction potential of 1.00 M H^+. Under standard conditions this value would be -0.44 V, indicating a lower tendency of 1.00 M Fe^{2+} to be reduced than 1.00 M H^+. For this reason, in this system, H^+ would be reduced, $H^+ + 2e^- \rightarrow H_2$, while Fe would be forced to oxidize, $Fe \rightarrow Fe^{2+} + 2e^-$. As a consequence, iron corrodes in the galvanic cell.

The potential measured using the experimental setup shown in Figure 1.3 is a reversible equilibrium potential of the half-cell electrode reaction, $Fe^{2+} + 2e^- \Leftrightarrow$ Fe. This electrode potential can be related to the fundamental electrode thermodynamic equation, the *Nernst equation* [equation (1.1)], which gives an electrode potential for any reversible half-cell reaction, $Ox + ne- \Leftrightarrow Red$, as a function of

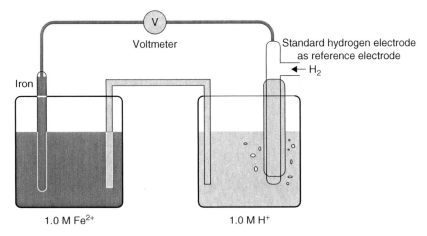

Figure 1.3   Electrode potential measurement.

nonstandard ionic concentrations:

$$E = E^\circ + \frac{RT}{nF} \ln \frac{[\text{Ox}]}{[\text{Red}]} \tag{1.1}$$

where E is the nonstandard half-cell reduction potential of interest, E° the standard half-cell reduction potential, [Ox] the concentration of oxidant Ox, [Red] the concentration of reductant Red, F is Faraday's constant, R the universal gas constant, T the absolute temperature, and n the number of electrons transferred in the reaction.

For an iron electrode reaction, $Fe^{2+} + 2e^- \Leftrightarrow Fe$, the Nernst equation gives the relationship between its reduction potential and the Fe^{2+} concentration:

$$E = E^\circ + \frac{RT}{2F} \ln[Fe^{2+}] \tag{1.2}$$

The potential measurement method illustrated in Figure 1.3 is valid only under the assumption that the iron electrode–solution interface is homogeneous, and therefore that the iron electrode potential measured can be considered to be equal to the potential at any location of the iron electrode surface. This homogeneous electrode assumption is also essential for the relationship in equation (1.2) because the Nernst equation is valid only when the concentration of Fe^{2+} is distributed uniformly over the electrode–solution interface, and therefore the electrode potential calculated can be considered to be equal to the potential at any location of the iron electrode surface.

Classical theory describing the kinetics of Faradaic electron transfer processes in dynamic electrochemistry is also based on this homogeneous electrode–solution interface model. The most fundamental electrode reaction kinetic relationship, the

Bulter–Volmer equation [1,6,7], is used to describe the kinetics of anodic and cathodic half-cell reactions:

$$i = i_o \left[\exp\left(\frac{\alpha n F}{RT} \eta \right) - \exp\left(\frac{-\beta n F}{RT} \eta \right) \right] \tag{1.3}$$

where η is the overpotential ($= E - E_{eq}$), the amount by which a potential deviates from equilibrium potential; i the measured or applied current density; i_o the exchange current density, F is Faraday's constant, R the universal gas constant, T the absolute temperature, n the number of electrons transferred in the reaction, and α and β are anodic and cathodic charge transfer coefficients that are related to the potential drop through the electrochemical double layer.

The Bulter–Volmer equation suggests that the kinetics of an electrochemical reaction depends on various factors, including overpotential and exchange current density. Positive overpotentials enhance oxidation reactions on the electrode, whereas negative overpotentials induce reduction reactions to occur. For a given overpotential, the magnitude of reaction current flow is affected particularly by the charge transfer rate between solution species and the electrode. The charge transfer rate is a key parameter indicating the kinetics of an electrochemical reaction. It is usually measured by the exchange current density, i_o, using the electrochemical polarization measurement setup shown in Figure 1.4.

In polarization measurements, a potentiostat is used to perturb the working electrode by injecting or extracting electrons to or from the electrode, causing the electrode potential to change from its equilibrium potential E_{eq} to a new electrode potential, E. The overpotential, $\eta = E - E_{eq}$, is a response to the applied current, which can be adjusted to any value desired. The electrode potential can be measured using a voltmeter and a reference electrode, while the applied current can

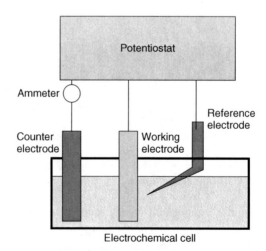

Figure 1.4 Experimental setup for polarization measurement.

be measured using a zero-resistance ammeter. The measured potential vs. current (or log i) data are usually presented as polarization curves. Experimental polarization curves contain information on electrode reaction kinetics and factors affecting electrochemical processes and mechanisms.

Apparently, the Butler–Volmer equation and the polarization measurement are also based on the assumption that an electrode–solution interface is homogeneous and therefore that electrode potential and currents can be considered to be distributed uniformly over the electrode surface. It should also be noted that the Butler–Volmer equation describes only an idealized active controlled electrode. That is, the electrode kinetics limits the current flow; it is not applicable to situations where electrochemical kinetics are controlled by mass transport of reactant molecules or ions from the bulk electrolyte. More detailed descriptions of traditional electrochemical thermodynamic and kinetic theories and methods may be found in books by Bard and Faulkner [1], Bockris et al. [2], and Mansfeld [4].

1.2 MIXED ELECTRODES AND UNIFORM CORROSION MODELS

Unlike the ideal electrode–solution interface shown in Figure 1.2 and the ideal half-cells shown in Figure 1.3, practical electrode processes such as aqueous metal corrosion involve at least two electrode–solution interfaces and two half-cell reactions occurring simultaneously at the anodic and cathodic half-cells. A typical example is the galvanic corrosion cell shown in Figure 1.1a, which involves an anodic metal dissolution reaction (Fe $\rightarrow$ Fe^{2+} + 2e$^-$) and a cathodic hydrogen evolution reaction (2H$^+$ + 2e$^-$ $\rightarrow$ H$_2$) occurring over two separate electrode–solution interfaces. Other examples of galvanic cells include cells made up of electrodes of two dissimilar metals that are connected electrically, and cells containing electrodes of the same electrode material exposed to different oxygen concentrations [3].

In most practical cases, aqueous corrosion reactions occur over a single piece of metal surface rather than over two separate electrodes in a galvanic cell. Such a corroding metal surface is often referred to as a *mixed electrode* [8] since several different redox reactions with different kinetics and reduction potentials occur simultaneously over the same electrode surface. A microelectrochemical cell model is proposed to explain the uniform corrosion phenomenon, based on the following assumptions:

1. Many tiny microscopic anodes and cathodes form on a single piece of corroding metal surface. Anodes and cathodes are distributed randomly over areas of the metal surface with different electrochemical potentials.
2. Individual half-reactions occur in these microscopically separated half-cells, causing an anode to corrode and an electron transfer to the cathode through an internal electrical circuit. Ions flow through a conducting solution on the metal surface.
3. Anode and cathode locations change dynamically, and a given area on a metal surface acts as both an anode and a cathode over any extended period

of time. The averaging effect of these shifting local electrochemical cells results in a rather uniform attack and general loss of material and roughening of the surface.

Figure 1.5 shows magnified representations of typical corroding iron surfaces exposed to three different environmental conditions. Figure 1.5a shows a typical iron electrochemical corrosion cell in acidic media where hydrogen evolution is the predominate cathodic reaction. Figure 1.5b shows a magnified corrosion cell under neutral pH or alkaline conditions where oxygen reduction ($O_2 + 2H_2O + 4e^- \rightarrow 4OH^-$) is the predominate cathodic reaction. Figure 1.5c shows a more complex corrosion system, where more than one reduction reaction occurs simultaneously over an electrode surface. In these microscopic corrosion cells the major corrosion anodic reaction is determined by which species is most easily oxidized, while the major cathodic reaction is the one with the highest reduction potential (i.e., the one with the greater tendency to undergo reduction). In the corrosion system shown in Figure 1.5a, the oxidation of iron has the lowest reduction potential and is therefore forced to undergo oxidation, $Fe \rightarrow Fe^{2+} + 2e^-$ ($E^0 = -0.44$ V vs. SHE, iron corrosion), and the reduction of hydrogen is the major cathodic reaction because it is the most easily reduced species in an oxygen-free environment, $2H^+ + 2e^- \rightarrow H_2$ ($E^0 = 0$, hydrogen evolution). If other oxidizing agents, such as oxygen and cupric ion (Cu^{2+}), exist in the corrosion environment, as shown in Figure 1.5c, there will be competing cathodic reactions, $O_2 + 4H^+ + 4e^- \rightarrow 2H_2O$ ($E^0 = +1.23$, oxygen reduction) and $Cu^{2+} + 2e^- \rightarrow Cu$ ($E^0 = +0.34$, copper plating).

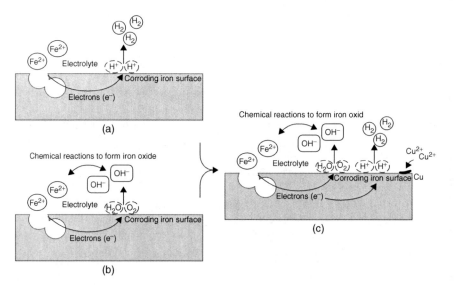

Figure 1.5 Magnified representation of electrochemical corrosion cells over iron electrodes exposed to various environmental conditions.

More complex mixed electrode–solution interfacial models are often needed to describe practical industrial corrosion processes. Figure 1.6 illustrates a mixed electrode–solution interfacial model illustrating corrosion of iron in a CO_2-containing aqueous solution under ambient temperature and atmospheric pressure conditions. Corrosion of iron in a CO_2-containing aqueous environment is a typical electrochemical corrosion case that commonly occurs in oil and gas pipelines. CO_2 is a naturally occurring component in many oil and gas fields, where CO_2 is associated with water, oil, and gas production. In aqueous environments, CO_2 dissolves and forms carbonic acid, leading to various forms of steel pipeline corrosion. It is known that CO_2 is significantly more corrosive than normal weak acid and that at a given pH, more corrosion of steel is caused by aqueous CO_2 solution than by hydrochloric acid [9]. This experimental fact suggests that hydrogen ion is unlikely to be the major corrosive species in CO_2 corrosion (i.e., the reduction of H^+ is unlikely to be the main cathodic reduction reaction in CO_2 corrosion). De Waard and Milliams concluded that the cathodic hydrogen evolution in CO_2 corrosion proceeds in a "catalytic" manner by direct reduction of undissociated adsorbed carbonic acid [10]. This mechanism is now generally accepted as the explanation for the strong corrosivity of carbonic acid. The detailed process of the cathodic reduction reaction of CO_2 corrosion was reported by Schmitt [11], and the cathodic reaction mechanism proposed is shown in Figure 1.6.

As shown, in CO_2 corrosion processes, anode and cathode locations change from period 1 to period 2. A given area on a metal surface would act as both an

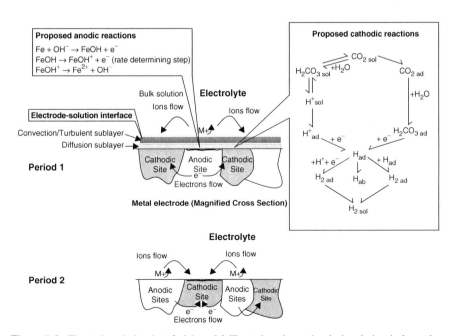

Figure 1.6   Electrode–solution interfacial model illustrating electrochemical and chemical reactions leading to uniform iron corrosion in CO_2-containing aqueous solution.

anode and a cathode over any extended period of time, and the averaging effect of these shifting local reactions results in a rather uniform dissolution of metal surfaces; therefore, this corrosion model would lead to uniform corrosion.

If a more careful investigation of the electrode–solution interface is carried out, it can be found that the uniform CO_2 corrosion of iron also involves transport processes occurring simultaneously over an electrode–solution interface. Since electrochemical and chemical reactions would cause the concentration of certain species in the solution (e.g., Fe^{2+}) to increase whereas others will be depleted (e.g., H^+), the established concentration gradients will lead to the movement of reactants and products toward and away from the electrode surface. If the transportation of reaction species due to diffusion and convection processes is unable to sustain the speed of the electrochemical reactions, the concentration of species at the electrode–solution interface can become very different from those in the bulk solution [12]. In this case, the electrode–solution model needs to accommodate a two-way coupling between the electrochemical corrosion processes at the metal surface and the diffusion and electromigration processes in the adjacent solution layer. Many corrosion models described in the literature have been developed to illustrate various forms of uniform or general corrosion in more complex environmental conditions.

1.3 MIXED POTENTIAL THEORY AND ELECTROCHEMICAL CORROSION MEASUREMENT

Wagner–Truad proposed mixed potential theory to explain the operation of mixed electrode cells operating at a mixed potential [8]. According to this theory, any electrochemical reaction can be divided algebraically into separate oxidation and reduction reactions with no net accumulation of electrical charge. In the absence of an externally applied potential, the oxidation of a metal and the reduction of species in solution occur simultaneously at a metal–electrolyte interface. Under these circumstances, the net measurable current is zero and the corroding metal is charge neutral, with no net accumulation of charge. For metals, the electrochemical potential of a metal at the anodic site is assumed equal to that at the cathodic site, due to its very low resistance [8].

Corrosion is a typical mixed electrode process operating at a mixed potential: the corrosion potential. The corrosion potential can be measured using an experimental setup similar to that shown in Figure 1.3, by recording the potential difference between a corroding electrode where both anodic and cathodic reactions occur, and a stable reference electrode. The corrosion potential is commonly used in conjunction with an *E–pH diagram* (often referred to as a *Pourbaix diagram*) as an indicator of the corrosion thermodynamic status to predict if corrosion will occur, to estimate the composition of corrosion products, and to predict environmental changes that would prevent or reduce corrosion attack. The *E*–pH diagram is a graphical representation of the thermodynamics of common electrochemical and chemical equilibria between metal and water, indicating thermodynamically stable phases as a function of electrode potential and pH [13]. An E–pH diagram can be constructed through application of the Nernst equation to each of these electrode reactions as a function of pH. Many E–pH diagrams have already been constructed for common material-environment systems by corrosion scientists, including those in Pourbaix's laboratory [14]. In many cases, an E–pH diagram can be found from the literature for a particular material environment, although we may need to construct E–pH diagrams for less common systems. Figure 1.7 shows a typical version of the E–pH diagram for an iron–water system at ambient temperature.

An E–pH diagram visualizes the thermodynamics of corrosion processes and gives information about a metal surface, whether it is in a region of immunity where the tendency for corrosion is nil, in a region where the tendency for corrosion is high, or in a region where the tendency for corrosion may still exist but where there is also a tendency and possibility for a protective or passive film to exist. However, the thermodynamically derived Pourbaix diagram only provides information on corrosion tendency. Like any thermodynamic quantity, the mixed potential value on its own does not provide information on the rate of corrosion. The determination of corrosion rate requires measurement of the kinetics of the corrosion electrochemical process.

The rate of a corrosion reaction (e.g., Fe $\rightarrow$ Fe^{2+} + 2e$^-$), could be determined if we are able to measure the flows of electrons in the metallic phase or ions in the aqueous phase because the corrosion current, i_{corr}, should be the sum of electron flows. However, unfortunately, corrosion electron flows could not be measured easily from corroding surfaces because we are unable to measure directly electrons

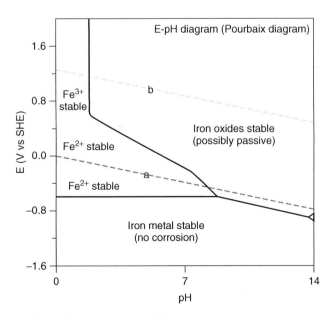

Figure 1.7 Simplified E–pH diagram for an iron–water system.

flowing between numerous minianodes and minicathodes located on the same electrode surface. We need to find an indirect way to determine i_{corr} from a corroding electrode surface.

A method of analyzing the kinetics of a mixed electrode under dynamic corrosion is a graphical representation of the kinetics of a mixed electrode (often referred to as an *Evans diagram*) [5,14]. Figure 1.8 shows an Evans diagram for iron electrode corrosion in acid. A mixed potential, the corrosion potential E_{corr}, is achieved through shifting the potentials of both anodic and cathodic reactions by a corrosion current, i_{corr}. This diagram illustrates a nonequilibrium system where dynamic reactions produce and maintain a corrosion current.

An Evans diagram is a very useful tool for analyzing corrosion kinetics and for predicting factors that may affect corrosion reactions; however, it should be noted that an Evans diagram is an imaginative and simplified illustration of electrode "internal" polarization over a corroding metal surface. This diagram is not directly measurable experimentally, and thus it is unable to reveal i_{corr} from a corroding electrode surface.

Pioneers in this field performed numerous experimental and theoretical analyses to find the "missing link" between measurable data from a corroding electrode and i_{corr} values. One idea of how to determine i_{corr} experimentally is to apply a perturbation signal to the electrode using a polarization experimental setup similar to that shown in Figure 1.4 [14]. In polarization measurements, a potentiostat is used to polarize an electrode from its steady-state mixed potential E_{corr} (note: not E_{eq})

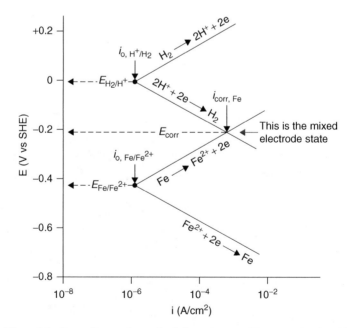

Figure 1.8 Evans diagram for a mixed electrode state of iron corrosion in acid.

to a new electrode potential, E. The overpotential, $\eta = E - E_{corr}$, is a response to the current applied, which can be adjusted to any value desired. The measured potential vs. current (or log i) data are usually presented as *polarization curves*. Figure 1.9 illustrates experimental polarization curves that are measurable from iron corrosion in acid, together with a comparison of an experimental polarization diagram to an Evans diagram.

The Evans diagram and the polarization curve can both be used to describe mixed electrode processes leading to metal corrosion. They are different but related. The Evans diagrams in Figure 1.9 are simplified plots of anodic and cathodic reaction curves of an electrochemical system, while the polarization curves describe the potential current density behavior obtained experimentally by applying an external potential or current. It is clear that the anodic and cathodic polarization curves merge with the Evans diagram under an area of relatively high polarization (usually referred to as the *Tafel region*). This suggests that the corrosion potential E_{corr} and corrosion current i_{corr} could be determined by extrapolation of the polarization curve. This method, often referred to as the *Tafel method*, was discovered experimentally by a pioneer in this field [3,14].

The fundamental formula describing the kinetics of an electrochemical reaction, the Butler–Volmer formulation [equation (1.3)], can be applied in conjunction with mixed-potential theory to determine corrosion kinetics over a uniform electrode

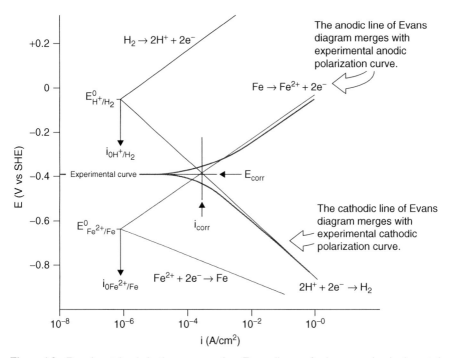

Figure 1.9 Experimental polarization curves and an Evans diagram for iron corrosion in deaerated acidic solution.

surface. If anodic and cathodic reactions over an electrode surface are activation controlled and the corrosion potential is far away from the equilibrium potentials of the individual anodic and cathodic reactions, the Butler–Volmer equation can be applied to a mixed electrode system:

$$i = i_{corr} \left[\exp\left(\frac{\alpha n F}{RT}\eta\right) - \exp\left(\frac{-\beta n' F}{RT}\eta\right) \right] \quad (1.4)$$

where η is the overpotential ($\eta = E - E_{corr}$), i the measured current density, i_{corr} the corrosion current density, F is Faraday's constant, R the universal gas constant, T the absolute temperature, n and n' the number of electrons transferred in the anodic and cathodic reactions, and α and β are coefficients related to the potential drop through the electrochemical double layer.

Equation (1.4) is a fundamental formula for measuring corrosion reaction kinetics. In its practical application, this equation is often simplified. The two most commonly used simplified forms of the equation are the Tafel equation and the Stern–Geary equation. The Tafel equation can be deduced from equation (1.4) for sufficiently high values of the applied potential. For anodic polarization, when $\eta \gg RT/\beta n' F$, equation (1.4) can be simplified as

$$i = i_{corr} \left[\exp\left(\frac{\alpha n F}{RT}\eta\right) \right] \quad (1.5)$$

that is,

$$\eta = -\frac{2.3RT}{\alpha F} \log i_{corr} + \frac{2.3RT}{\alpha F} \log i \quad (1.6)$$

or, for cathodic polarization, when $-\eta >> RT/\alpha n F$, equation (1.4) can be simplified as

$$i = i_{corr} \left[\exp\left(\frac{-\beta n' F}{RT}\eta\right) \right] \quad (1.7)$$

that is,

$$-\eta = -\frac{2.3RT}{\beta F} \log i_{corr} + \frac{2.3RT}{\beta F} \log i \quad (1.8)$$

Equations (1.6) and (1.8) have the form of the Tafel equation:

$$|\eta| = a + b \log i \quad (1.9)$$

where a and b are Tafel constants: $a = -(2.3RT/\alpha F) \log i_{corr}$ and $b = (2.3RT/\alpha F)$ for anodic polarization, or $a = -(2.3RT/\beta F) \log i_{corr}$ and $b = (2.3RT/\beta F)$ for cathodic polarization.

Equation (1.9) provides a theoretical background to the Tafel method (also referred to as the *Tafel line extrapolation method*) that was discovered experimentally [3,14]. According to the Tafel equation (1.9), the value of either the anodic or the cathodic current at the intersection is i_{corr}. This suggests that corrosion current i_{corr} can be determined by extrapolating the linear portions of the Tafel plot back to their intersection, where the overpotential ($\eta = E - E_{corr}$) is zero. Figure 1.10 illustrates determination of the corrosion current, i_{corr}, using the Tafel line extrapolation method. The Tafel extrapolation method is often used to determine corrosion parameters, including corrosion current and Tafel slopes; however, the Tafel equation is true only for relatively high overpotentials, 100 mV or higher, and thus use of this method requires application of a large polarization voltage, which could cause irreversible damage to the electrode surface. The Tafel method is therefore not recommended for continuous corrosion rate measurement and thus is of only limited value for corrosion rate monitoring.

Stern and Geary simplified the Butler–Volmer formula for low polarization (≤ 10 mV) and developed an equation that can be used to determine the corrosion rate without significantly damaging the electrode surface [15]:

$$i_{corr} = \frac{b_a b_c}{2.3(b_a + b_c)} \frac{1}{R_p} \tag{1.10}$$

where i_{corr} is the corrosion current density, b_a and b_c are anodic and cathodic Tafel slopes, and R_p is the polarization resistance, which is defined as the tangent of a

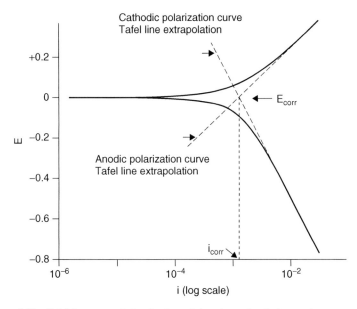

Figure 1.10   Tafel line extrapolation for determining electrochemical corrosion parameters.

polarization curve at the corrosion potential:

$$R_p = \left(\frac{d\eta_c}{di_c}\right)_{\eta_c \to 0} = \left(\frac{d\eta_a}{di_a}\right)_{\eta_a \to 0} \tag{1.11}$$

The Stern–Geary equation (1.10) provides a new method, usually referred as the *linear polarization resistance* (LPR) *method*, for corrosion current i_{corr} determination [16]. When measuring the corrosion rate using this method, only a small polarization voltage (normally, ± 10 mV) is applied to a freely corroding electrode, and the resulting "linear" current response is measured. The polarization resistance is the ratio of the applied perturbation potential and the resulting current response in equation (1.11). According to the Stern–Geary equation (1.10), this "resistance" is inversely related to the uniform corrosion current density i_{corr}. Figure 1.11 illustrates the LPR method for determining R_p. The b_a and b_c values are normally determined by a Tafel curve or weight-loss measurements. They can also be estimated by linear polarization measurement by carrying out a nonlinear least-squares fit of the dc linear polarization data to the Stern–Geary equation [16,17].

Compared to techniques using very large anodic or cathodic overpotentials, such as the Tafel method, the Stern–Geary method has advantages because it is essentially a nondestructive technique. It has become probably the most popular electrochemical method for determining the corrosion current, i_{corr}, which can be converted to a more convenient corrosion weight-loss value:

$$\text{weight-loss(g/cm}^2) = \int_0^T \frac{i_{corr}(t)E_w}{96,500s}dt \tag{1.12}$$

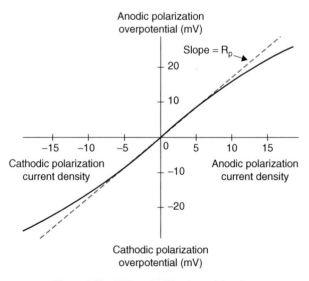

Figure 1.11 LPR method for determining R_p.

where $i_{corr}(t)$ is the corrosion current density at time t (in A/cm^2), s the electrode surface area (in cm^2), T the experimental duration (in seconds), and E_w the equivalent weight of electrode material (27.92 for iron).

In practice, electrochemical measurements are carried out regularly with interval δt. For simplicity, the corrosion rate during the interval of each measurement could be taken as a constant. So equation (1.12) can be rewritten as

$$\text{weight loss (g/cm}^2) = \sum_{n=1}^{k} \frac{i_{corr}(n)E_w}{96,500s}\delta_t(n) \tag{1.13}$$

where $\delta_t(n)$ is the test interval (in seconds) during which the corrosion rate is $i_{corr}(n)$. n is the series number of tests and k is the total number of tests during the entire experiment. Combining equations (1.10) and (1.13), for a mild steel electrode the corrosion weight loss can be calculated from R_p by

$$\text{weight loss (g/cm}^2) = 1.256 \times 10^{-4} \sum_{n=1}^{k} \left[\frac{b_a b_c \delta_t}{(b_a + b_c)s R_p} \right]_n \tag{1.14}$$

However, it should be noted that under some circumstances the LPR method has been found to be unreliable, and large errors can occur in LPR corrosion rate measurements [18–20]. One of the reasons for these errors is that the experimental R_p value contains contributions from ohmic resistances, such as the solution resistance between the Luggin capillary and the test electrode, surface scale resistance and inhibitor film resistance, and so on. Although methods such as the positive feedback technique and the interrupter technique have been developed and employed to remove the effects of ohmic resistances, a detailed study of these techniques has shown that they have limitations and are very difficult to apply [19]. For these reasons, the LPR technique is generally not applicable to systems where there is large solution resistance, surface scale resistance, or inhibitor film resistance.

EIS has been developed as an alternative technique for polarization resistance measurement [21]. EIS is considered to be a more reliable technique because it is supposed to be able to separate different corrosion electrochemical processes and thus be able to eliminate measurement errors due to solution resistance and surface film resistance. EIS is also employed in the investigation of corrosion coating and inhibition through obtaining information about electrode–inhibitor film interfacial characteristics and integrity [4,22]. Corrosion measurement using EIS has already been discussed by many authors, in particular Mansfeld [4].

Electrochemical noise analysis (ENA) is another method that has been used to determine polarization resistance by measuring the electrochemical noise resistance [23]. Noise resistance is the ratio of the standard deviation of voltage noise and the standard deviation of current noise. It has been found to be similar (or equivalent) to the charge transfer resistance or polarization resistance and thus can be used to calculate corrosion rates [20]. Noise resistance measurements use only simple instruments and do not apply perturbation to the test system by an externally imposed polarization, which would lead to inevitable changes in such system-specific properties as the surface structure and roughness, and the sorption processes of inhibitors. More detailed descriptions of the theory, measurement, and analysis of the noise resistance technique may be found in a book [24] and several articles [25–28].

To understand the advantages and disadvantages of major corrosion measurement techniques, including weight-loss measurement, LPR, ENA, and EIS, these techniques have been applied simultaneously to measure uniform CO_2 corrosion with and without the influence of surface scales and inhibitor films [20,28]. In simple low-resistance corrosion environments such as a bare electrode–coupon in an electrochemical cell containing 640 mL of 3% NaCl brine with CO_2 sparging at 50°C all techniques yielded similar corrosion rates (Figure 1.12), [20]. This result suggests that all these techniques can be used to measure corrosion rates in low-solution-resistance systems.

When a bare electrode was put in 3% NaCl brine in the presence of 50 ppm of the inhibitor imidazoline with CO_2 sparging at 30°C polarization resistance values from LPR, EIS, and ENA measurements generally correlate well (Figure 1.13). This result suggests that these techniques can be used effectively to measure corrosion rates on electrode surfaces with inhibitor adsorption [20]. However, with the formation of a high-resistance inhibitor film on an electrode surface, the LPR technique was found to lose accuracy. In an experiment, a thick commercial batch treatment inhibitor was formed on steel electrode–coupon surfaces, and major differences between the weight-loss and electrochemical methods were observed [20]. Weight-loss and electrochemical measurement results have a difference of more than 500-fold. It is suggested that the reliability of LPR should be checked

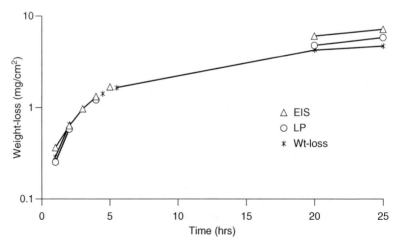

Figure 1.12 Comparison of weight loss measured using weight-loss coupons, EIS, and LP of a bare electrode and coupons in 640 mL of 3% NaCl brine with CO_2 sparging at 50°C. (From [20].)

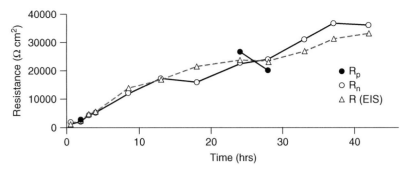

Figure 1.13 Comparison of R_p (LPR), R (EIS), and R_n (ENA) measured during inhibitor film formation. (From [20].)

by EIS and weight-loss measurements if the corrosion environment has a high solution resistance or if a dielectrical surface film is present [20].

Each electrochemical technique has advantages and limitations; therefore, different techniques are often applied in a synergistic manner. For example, corrosion potential measurements are used to investigate the effects of corrosion inhibitors on corrosion thermodynamics, while LPR, EIS, and ENA are used to monitor changes in electrode processes kinetics. Care should be taken, however; these techniques are accurate only under several fundamental assumptions [19], and in principle these techniques are applicable only to uniform corrosion systems.

1.4 ELECTROCHEMICAL IMPEDANCE INVESTIGATION OF AN ELECTRODE–SOLUTION INTERFACE

Electrode impedance is one of the most important quantities that can be measured in electrochemistry and corrosion science by purely electrical means. Since Epelboin and co-workers introduced ac impedance techniques in the 1960s [21], EIS has become an important technique that has broadened the range of corrosion phenomena that can be studied using electrochemical techniques [4,22]. Although there are pitfalls [29] and critical issues [30] associated with its application, EIS offers distinct advantages over dc techniques. Two main areas of application of the EIS technique are rapid estimates of a wide range of corrosion rates, and practical insights into corrosion mechanisms. More detailed explanations of the theory, analysis, and application of the EIS can be found in articles authored by Mansfeld and others [4,21,22, 29–31].

The power of EIS lies in the fact that it is capable of accessing relaxation phenomena with relaxation times varying over many orders of magnitude (e.g., 10^6 to 10^{-4} Hz), permitting a wide range of interfacial processes to be investigated, including interfacial reactions over corroding electrode surfaces [31]. The interpretation of EIS data allows one to determine the electrochemical parameters of the electrode–solution interface. For example, EIS can provide information on electrode interfacial structure, electrode capacitance, and charge transfer kinetics. EIS Nyquist plots are used as a basis for modeling the electrode–solution interface as an electrical circuit. This is particularly useful for the study of an electrode–solution interface in the presence of an adsorbed surface film or a dielectric organic coating [32]. Characterization of the adsorption, desorption, and film formation on an electrode surface may be studied by determining its surface resistance and capacitance.

Interpretation of experimental EIS data requires suitable models for simulating impedance behavior and in developing fitting programs for analyzing EIS data. A typical impedance Nyquist plot of an electrode with a surface coating present is shown in Figure 1.14a. This impedance characteristic is common in the organic coating–metal electrode and inhibitor film–metal electrode corrosion systems. The two semicircles suggest that this is a multiple-time-constant system. A semicircle at the lower frequency is usually due to the corrosion electrochemical process (i.e., the charge transfer process) [32,33]. A semicircle in the high-frequency range would be due to the inhibitor film because a surface dielectric film normally has a small time constant [34]. This has been verified experimentally by carrying out weight-loss and EIS measurements simultaneously over a 70-hour inhibitor test period [20]. In the experiment the average weight loss of corrosion coupons was 0.55 mg/cm^2, while the weight loss calculated from low-frequency-semicircle EIS data using equation (1.14) was about 0.60 mg/cm^2 and that calculated from high-frequency-semicircle EIS data was about 2.3 mg/cm^2. Obviously, the low-frequency semicircle correlates much better with corrosion coupon test results. This experiment confirms that the lower-frequency semicircle was due to a charge transfer process [20].

However, in practical systems, there is difficulty in the analysis of impedance spectra. The shape of the Nyquist plot appears as two distinct semicircles only

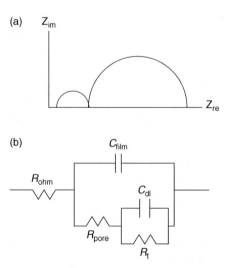

Figure 1.14 (a) The impedance Nyquist plot of an electrode filmed with a non-conducting surface film. (b) Equivalent circuit for the electrode.

if both of the following criteria are met [35]: $0.2 \leq R_t/R_{film} \leq 5$; $\tau_{dl}/\tau_{film} \geq 20$, where R_t is the charge transfer resistance, R_{film} the coating film resistance, τ_{dl} the time constant for the double layer, and τ_{film} the time constant for the coating film. If these relationships are not met, the two semicircles could overlap, causing difficulties in determining the components of the equivalent electrical circuit.

The impedance characteristics of this electrode surface can be simulated by an electrical circuit proposed for coated metal electrodes [36], as shown in Figure 1.14b. In the electrical circuit, R_{ohm} is the solution resistance, R_{pore} the resistance of the surface film in the pore area, C_{film} the capacitance of the surface film, R_t the charge transfer resistance, and C_{dl} the double-layer capacitance. Computer programs such as the Boukamp program [37], which uses a nonlinear least-squares fit (NLLS-fit) and simulation methods to fit experimental EIS data, are useful in the analysis of complicated EIS spectra. However, it should also be noted that one of the pitfalls in the EIS technique is the ambiguity inherent in selecting an appropriate model for interpreting impedance spectra [30]. It was pointed out that a number of plausible models can be found to fit a given EIS result. The true situation of the system may be different from the simplest model suggested, which would normally be used to simulate the practical system. Therefore, model identification requires additional information, and improvement in model identification capabilities is especially important [30]. The development of proper models often requires prior knowledge of the system-specific corrosion behavior and electrode–solution interfacial structure. Like other electrochemical techniques, EIS cannot be regarded as being a "stand-alone" technique and should be employed only very cautiously.

In a typical experiment, EIS plots such as those in Figure 1.15 were recorded from a mild steel electrode after being exposed to CO_2-saturated 3% NaCl brine for various periods [38]. As shown in Figure 1.15 a, electrode impedance increased significantly after 24 hours' exposure to 64 mL of CO_2-saturated 3% NaCl brine at 70°C, indicating the formation of a protective corrosion product scale on the mild steel electrode surface. The appearance of a new phase-angle shift in the Bode phase plots in Figure 1.15b indicates a scale formed on electrodes after 48 hours' exposure. Obviously, Nyquist and Bode phase plots can be used to provide useful information on protective abilities and electrochemical properties of corrosion scales. The CO_2 corrosion product scale is known to play an important role in the mechanism of, and protection against, CO_2 corrosion. In fact, the formation of corrosion product scale is linked to almost all of the most important issues in CO_2 corrosion science. For example, scale is believed to be the key factor resulting in significant changes in CO_2 corrosion kinetics and the invalidity of the de Waard–Milliams equation [10]. It was found that the formation of scale can reduce initial corrosion rates by as much as three orders of magnitude [39]. With corrosion scale present, mass transfer to and from the metal surface, instead of cathodic hydrogen evolution, can become the corrosion-rate-controlling factor [11,39]. Corrosion product scale, with its nonuniform formation and localized destruction, is also supposed to be the key factor causing localized CO_2 corrosion [11,40]. EIS measurements confirmed that CO_2 corrosion scales formed at high temperature

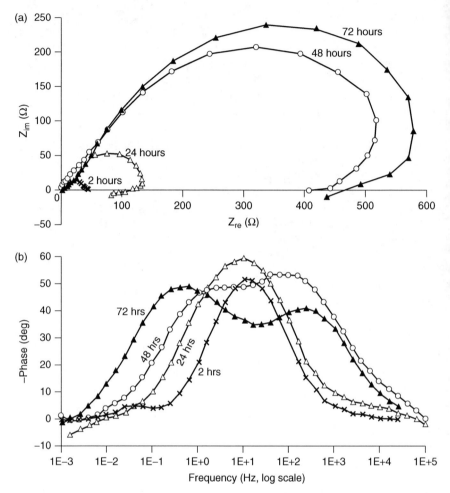

Figure 1.15   (a) Nyquist plots and (b) Bode phase plots recorded from prescaled mild steel electrodes that were exposed individually to 64 mL of CO_2-saturated 3% NaCl brine for 2, 24, 48, and 72 hours at 70 C. (From [38].)

and pressure provided better protection than those formed at low temperature and pressure. The level of protection of the scale formed at higher temperature and pressure increased with exposure time [38].

Corrosion inhibition plays a key role in CO_2 corrosion control in the oil and gas production industry. CO_2 corrosion inhibition is generally believed to be the result of physical adsorption and chemisorption, although they could also adsorb on the surface of iron carbonate scale and retard the mass transfer–dependent dissolution rate of the corrosion scale. In a typical experimental study, a mild steel electrode was exposed to a test solution containing 3% NaCl brine and 50 ppm of imidazoline, with continuous CO_2 sparging at 30°C. The electrode was kept stationary

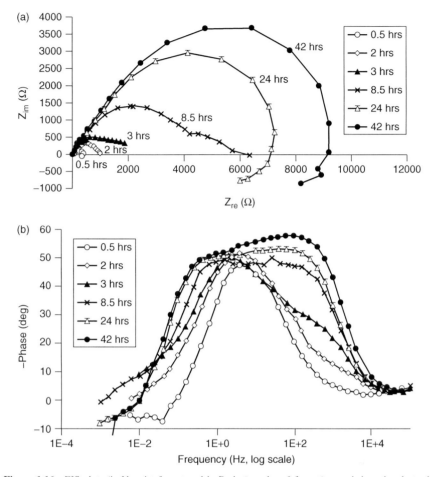

Figure 1.16 EIS plots (in Nyquist format and in Bode θ vs. log f format) recorded on the electrode after filming with the inhibitor imidazoline for various periods. (From [41].)

during the inhibitor filming process. Figure 1.16 shows EIS Nyquist and Bode plots that were recorded during the 42 hours of inhibitor-film formation [41]. As shown in Figure 1.16a, the diameter of Nyquist semicircles increased on a continuous basis, suggesting that the presence of inhibitor imidazoline greatly, but gradually, changed the corrosion kinetics on the electrode surface. Bode θ vs. log f plots prepared using the same experimental data as those used in the Nyquist plots are shown in Figure 1.16b. These plots show a new phase-angle shift in the higher frequency range and a continuous increase in the phase-angle shift with filming time. This new phase shift suggests that the formation of inhibitor film changed the electrode interfacial structure and resulted in an extra time constant. The continuous increase in the phase-angle shift apparently correlates with the inhibitor film growth.

Two peaks in the θ vs. log f plots mean that there are two major electrochemical processes on the electrode surface. In the Nyquist plots shown in Figure 1.16a, two semicircles appear, one of which is much smaller than the other and cannot be clearly recognized. The semicircle at the lower frequency, which shows peaks around 1 Hz in Figure 1.16b, is believed to be due to the corrosion electrochemical process [32,33] and has been confirmed using weight-loss data [20]. The semicircle in the high-frequency range would be due to the inhibitor film because a surface dielectric film normally has a small time constant and so has a phase-angle shift in the high-frequency range [22]. However, a simple semicircle fitting according to the model in Figure 1.14b is unable to fit the data points at intermediate frequencies in Figure 1.16a, suggesting the existence of some interfacial structures on the electrode surface which have intermediate time constants. The large phase-angle shift in the higher-frequency range in Figure 1.16b suggests a probable explanation: that the inhibitor film has a multilayered structure. An equivalent circuit in Figure 1.17 is suggested for an electrode with a four-layer inhibitor film [41]. In the equivalent circuit, R_{ohm} is the solution resistance, R_1 to R_4 the resistances of inhibitor layers, C_1 to C_4 the capacitances of inhibitor layers, R_t the charge transfer resistance, and C_{dl} the double-layer capacitance. Based on this multilayer interfacial model, the spectra can be fitted satisfactorily by a nonlinear least-squares fit (NLLS-fit) and simulation computer program [37]. Figure 1.18 shows the fitting of the typical Nyquist plot in Figure 1.16.

The electrochemical kinetics parameters of the electrode surface, including the resistances and capacitances of inhibitor layers (R_1 to R_4 and C_1 to C_4), electrochemical charge transfer resistance, and double-layer capacitance, can be deduced using the NLLS-fit and simulation program, and then the properties of the inhibitor film and electrochemical double layer can be defined. Table 1.1 shows the electrode surface electrochemical kinetics parameters during formation of the inhibitor film.

As shown in Figure 1.18 and Table 1.1, satisfactory simulation of the impedance characteristics of the inhibitor-filmed electrode surface by a four-layer model suggests that the inhibitor film should have a multilayered structure. The resistance

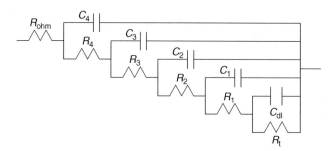

Figure 1.17 Equivalent circuit proposed for an electrode filmed with a four-layer nonconducting inhibitor film. (From [41].)

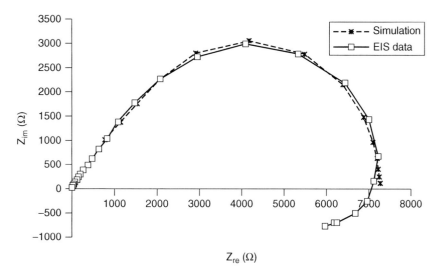

Figure 1.18   Computer simulation of a typical Nyquist plot using a four-layer model. (From [41].)

Table 1.1   Continuous Changes in the Resistances and Capacitances of Inhibitor Layers During Inhibitor-Film Formation

Filming (h)	$R_f(\Omega)$	$C_{dl}(\mu F)$	$R_1(\Omega)$	$C_1(\mu F)$	$R_2(\Omega)$	$C_2(\mu F)$	$R_3(\Omega)$	$C_3(\mu F)$	$R_4(\Omega)$	$C_4(\mu F)$
Before filming	15	1690								
0.5	122	951	159	193	63	119	10	83	2	22
2	207	7461	398	407	257	160	57	96	9	15
3	460	1678	649	211	205	103	49	39	20	12
4	483	4003	845	244	321	111	77	36	31	12
8.5	1119	1717	2082	169	600	63	200	22	62	11
13	2458	283	1736	98	565	34	178	13	61	11
18	3506	241	2110	91	653	34	198	13	63	10
24	3933	237	2312	89	719	33	221	13	69	11
28	4751	186	1619	57	424	19	102	10		
33	4741	223	2538	77	827	28	235	11	69	10
37	5020	220	2690	74	875	27	250	11	72	10
42	5246	213	2732	72	880	27	249	11	72	10

Source: [41].

values of each of the four inhibitor layers differs from the others, which suggests that each layer has a different inhibitor molecular density because a layer's resistance value is a reflection of the penetration of the layer by an electrolyte and so is related to the inhibitor molecular density in the layer. The first layer has the largest resistance, and thus it should have the densest inhibitor molecular structure. A possibility is that the inhibitor film consists of an inner layer that is likely to be an inhibitor–metal complex and several outer layers that are likely to be inhibitor layers with possible inhibitor molecular cross-linking. A physical model of the electrode–solution interface is suggested, as shown in Figure 1.19.

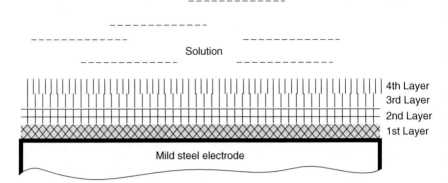

Figure 1.19 Physical model of an electrode surface filmed with a multilayer nonconducting inhibitor film (shows four layers only.) (From [41].)

After about 48 hours of filming, the electrodes were transferred from a solution containing 50 ppm of imidazoline into inhibitor-free brine, where the inhibitor film may deteriorate. The electrode was rotated at 1000 rpm to investigate inhibitor-film destruction by surface shear stress. Figure 1.20 shows EIS plots recorded on the electrode at different times. Table 1.2 shows electrochemical parameters calculated from EIS data. The continuous decrease in the diameter of Nyquist plots and the gradual disappearance of the high-frequency phase-angle shift (Figure 1.20) suggest that the inhibitor film was gradually removed by surface shear stress. The film destruction process was very slow, suggesting a strong interaction between the electrode and the inhibitor. On comparing Figure 1.20b with Figure 1.16b, very similar trends are seen, but in the opposite time series. This suggests that film destruction and film formation probably follow a "reversible" process.

In these experiments, EIS has been shown to be a valuable technique for studying the interfacial structure of an electrode in the presence of an inhibitor film, for understanding the mechanism of inhibitor-film formation and destruction, and for evaluating the film persistency of CO_2 corrosion inhibitors. EIS can be used to measure corrosion-related electrochemical parameters such as the resistances and capacitances of inhibitor layers, charge transfer resistance, and double-layer capacitance. These parameters can be used to analyze the inhibitor mechanism and to calculate the corrosion rate.

1.5 ELECTROCHEMICAL NOISE MONITORING OF RAPID ELECTRODE PROCESSES

Although EIS is a powerful technique for investigating the electrode–solution interfacial structures and the effects of corrosion inhibitors, it has limitations for monitoring rapid electrode interfacial changes on a continuous basis because EIS measurement takes quite a long time (often more than 1 hour). EIS is a steady-state

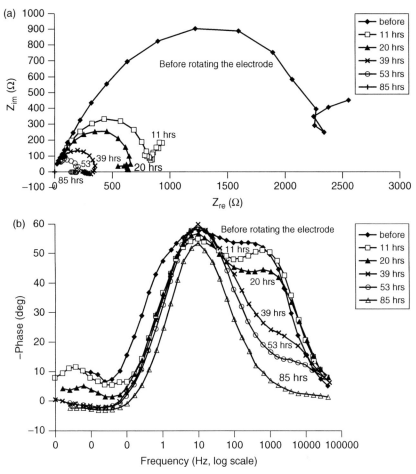

Figure 1.20 EIS plots before and after rotating an inhibitor-filmed electrode (1000 rpm) in an inhibitor-free 3% NaCl brine (From [41].)

technique that in principle cannot be used in corrosion systems where parameters such as potential are changing rapidly. For these reasons, EIS presents difficulties for researchers studying rapid corrosion electrochemical processes (e.g., corrosion inhibition processes immediately following inhibitor addition).

Electrochemical noise analysis (ENA) is a method that has been used to monitor dynamic changes in electrode interfacial processes by measuring electrochemical noise resistance (R_n) on a continuous basis. R_n is the ratio of the standard deviation of voltage noise and the standard deviation of current noise and is equivalent to the polarization resistance (R_p) [20,23–28]. In the typical experiment shown in Figure 1.21 a dual mild steel electrode with an inhibitor imidazoline film was quickly transferred into an inhibitor-free 3% NaCl solution with 1000-rpm stirring. The purpose was to monitor changes in electrode processes by recording the

Table 1.2 Continuous Changes in the Resistances and Capacitances of Inhibitor Layers After Rotating an Electrode at 1000 rpm in Fresh Brine

Testing h	$R_t(\Omega)$	$C_{dl}(\mu F)$	$R_1(\Omega)$	$C_1(\mu F)$	$R_2(\Omega)$	$C_2(\mu F)$	$R_3(\Omega)$	$C_3(\mu F)$	$R_4(\Omega)$	$C_4(\mu F)$	Corr. Rate (mm/yr)
Before stirring	1106	268	814	67	258	34	57	12	20	12	0.02
2	428	424	459	82	181	36	46	13	19	12	0.04
6	353	396	373	86	130	45	35	14	15	13	0.05
11	356	347	341	81	92	47	27	12	9	12	0.05
15	322	386	312	87	73	53	22	15	9	12	0.06
20	283	447	291	97	62	59	17	17	7	12	0.06
29	219	526	226	120	42	69	10	30	5	12	0.08
34	173	712	198	142	42	90	7	42	4	12	0.10
39	128	941	168	162	43	114	7	59	3	13	0.29
43	105	1122	142	186	34	139	5	65	2	13	0.35
53	68	1497	99	231	25	185	4	99	2	14	0.54
70	56	1609	82	261	23	213	3	111	1	14	0.66
75	38	2136	59	340	17	280	2	143	1	19	0.97
80	24	2894	41	445	13	336	2	230	0.5	33	1.54

Source: [41].

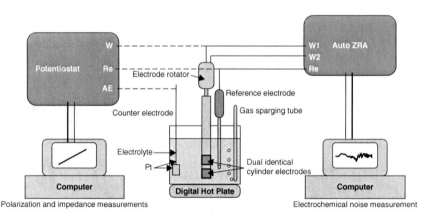

Figure 1.21 Experimental arrangement for electrochemical noise recording and for linear polarization measurement (From [28].)

electrode potential and current noise on a continuous basis, in conjunction with linear polarization (LP) measurements.

Figure 1.22 shows raw current noise recorded before and after electrode transference. After electrode transfer, the current noise amplitude apparently increased, which can be observed in the raw noise record and is shown more clearly in Figure 1.23, where the dc trend has been removed from the raw noise data. R_n decreased rapidly after electrode transfer, as shown in Figure 1.24, indicating rapid degradation of the inhibitor film. Although the R_n values in Figure 1.24 show fluctuations, the trend line appears to follow the breakdown process of the inhibitor

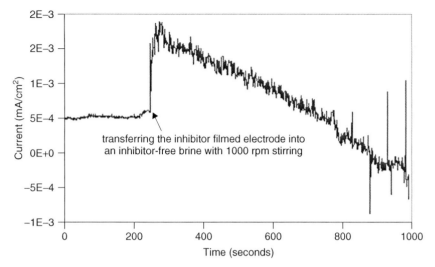

Figure 1.22 Original current noise records after transferring an inhibitor-filmed electrode into inhibitor-free brine with 1000-rpm stirring.

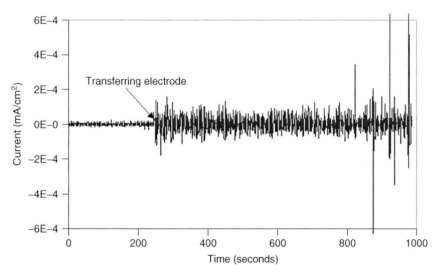

Figure 1.23 Current noise (after moving-average removal treatment) after transferring an inhibitor-filmed electrode into inhibitor-free brine with 1000-rpm stirring.

film. This result suggests that electrochemical resistance could be a convenient method of monitoring inhibitor-film breakdown and thus of evaluating inhibitor film persistency [28].

Figure 1.25 shows the inhibitor failure process monitored using different methods. Generally, ENA and LP showed similar trends during the inhibitor-film failure

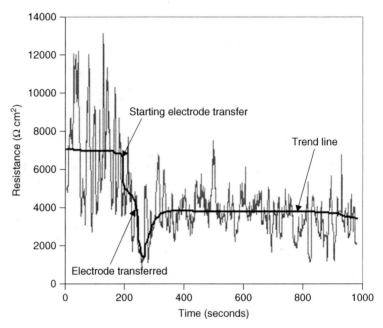

Figure 1.24 Change in electrochemical noise resistance (R_n) after transferring an inhibitor-filmed electrode into inhibitor-free 3% NaCl brine with 1000-rpm stirring.

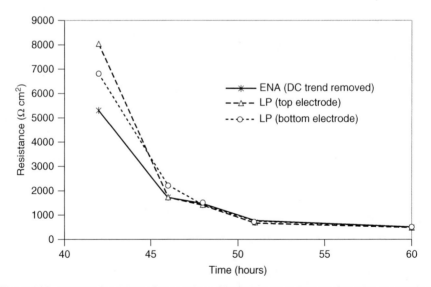

Figure 1.25 Inhibitor-film failure after exposing a filmed electrode to inhibitor-free brine and rotating at 1000 rpm. (From [28].)

process. Compared to LP, ENA has advantages in the continuous monitoring of inhibitor-film performance and in the convenient experimental arrangement. In another typical experiment, this continuous electrochemical noise resistance measurement method was used successfully to study the formation process of chromate conversion coatings on aluminum [42].

1.6   ISSUES AND DIFFICULTIES IN TRADITIONAL ELECTROCHEMICAL METHODS

Conventional electrochemical models and techniques based on the Nernst equation, the Bulter–Volmer formulation, and the mixed potential theory are powerful in understanding and measuring uniform corrosion processes, their thermodynamics, and their kinetics. However, it should be noted that these theories and methods are based fundamentally on a homogeneous or uniform mixed electrode–solution interface. The implicit assumption in formulating electrode process kinetics and testing them experimentally is that the current density is uniform and the potential varies only in the direction perpendicular to the surface. Indeed, experimental studies used to validate mixed potential theory employed mercury and zinc amalgam electrodes where a truly homogeneous surface is indeed likely to be achieved [8]. In principle, conventional electrochemical models and methods are not applicable to nonuniform electrode surfaces such as metal surfaces under pitting and crevice corrosion.

Figure 1.26 illustrates a typical metal surface under pitting corrosion where local chemical environments exist. When localized corrosion occurs, there is a distinct separation of the anodic and cathodic areas, where different electrochemical reactions occur. Electrons would travel continuously from the anode areas to the cathode areas through the electrode body, and at the same time, ions travel between anode and cathode areas through the electrolyte, resulting in rapid penetration of

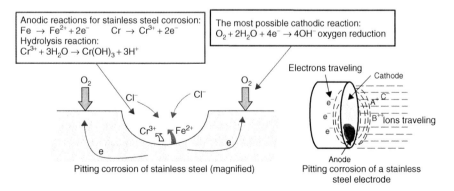

Figure 1.26   Electrode under pitting corrosion.

the metal at anodic areas and significant changes in surface chemistry and physical properties.

Nonuniform surface chemistry creates a major issue in electrochemical science and measurement. For example, conventional corrosion potential measurement using a macrodisk electrode, a reference electrode, and a voltmeter is applicable only to measuring the corrosion potential of uniform electrodes, where the potential at any location equals that of the entire electrode surface. If an electrode is heterogeneous, as shown in Figure 1.26, this method only detects a corrosion potential that is a mixture of contributions from many local potentials, none of which we can evaluate independently. This suggests that traditional electrochemistry has a major weakness—there is a "missing link" between electrochemical methods based on ideal uniform electrode models and practical nonuniform electrode processes.

Apparently, there is a need to bridge the major technological gap between conventional electrochemistry over uniform surfaces and heterogeneous electrochemistry over nonuniform surfaces [43–45]. Unfortunately, the modeling and measurement of nonuniform electrode processes are considered to be poorly understood aspects of solid-state physics, and this lack of understanding has contributed to inadequate experimental control over interfacial structure and reactivity [46].

It is well known that traditional electrochemical techniques are accurate only when they are applied to a study of the kinetics of uniform corrosion processes. In most practical electrochemical systems, such as locally corroding metals and surfaces under nonuniform electrodeposition or uneven electrodissolution, electrode surfaces are electrochemically nonuniform, and thus, in principle, traditional electrochemical techniques are not strictly valid, although this application is very common.

New methodologies are needed for the investigation of nonuniform electrode–solution interfacial structures as well as the thermodynamics and kinetics of heterogeneous electrode processes. Over the past decade, various innovative methods have been evolving, providing the methodological driving force for new investigations of the electrode–electrolyte interface. For example, scanning probe techniques that scan and detect local electrode potentials, galvanic currents, and local electrochemical impedances at the metal surface or metal–electrolyte interface have been developed and employed in localized corrosion research. An electrochemically integrated multielectrode array, the wire beam electrode, has also been developed for measuring, characterizing, and evaluating surface inhomogeneity, electrochemical heterogeneity, and localized corrosion [47].

REFERENCES

1. A. J. Bard and L. R. Faulkner, *Electrochemical Methods: Fundamentals and Applications*, 2nd ed., Wiley, New York, 2000.

2. J. O'M. Bockris, R. E. White, and B. E. Conway, (Eds.,) *Modern Aspects of Electrochemistry*, Vol. 31, Springer-Verlag, New York, 1998.

3. G. Fontana, *Corrosion Engineering*, 3rd ed., McGraw-Hill, New York, 1987.

4. F. Mansfeld, *Electrochemical Methods in Corrosion Testing*, ASM Handbook, Vol. 13A, ASM, Materials Park, OH, 2003, p. 445.

5. U. R. Evans, *An Introduction to Metallic Corrosion*, 3rd ed., Edward Arnold, London, 1981.

6. J. A. V. Butler, Studies in heterogeneous equilibria: I and II. Conditions at the boundary surface of crystalline solids and liquids, and the application of statistical mechanics; and The kinetic interpretation of the Nernst theory of electromotive force, *Transactions of the Faraday Society*, 19 (1924), 729, 734.

7. M. Volmer and T. Erdey-Gruz, The theory of hydrogen overvoltage, *Zeitschrift fuer Physikalische Chemie*, 150A (1930), 203.

8. F. Mansfeld, Classic paper in corrosion science and engineering with a perspective by F. Mansfeld, *Corrosion*, 62 (2006), 843.

9. G. W. Whitman, R. P. Russel, and V. J. Altieri, Effect of hydrogen-ion concentration on the submerged corrosion of steel, *Industrial and Engineering Chemistry*, 16 (1924), 665.

10. C. de Waard and D. E. Milliams, Carbonic acid corrosion of steel, *Corrosion*, 31 (1975), 177.

11. G. Schmitt, Fundamental aspects of CO_2 corrosion, in *Advances in CO_2 Corrosion*, R. H. Hausler and H. P. Godard, eds., NACE, Houston, TX, 1984, p. 10.

12. M. Nordsveen, S. Nesic, R. Nyborg, and A. Stangeland, A mechanistic model for carbon dioxide corrosion of mild steel in the presence of protective iron carbonate films: Part 1. Theory and verification, *Corrosion*, 59 (2003), 443.

13. M. Pourbaix, *Lectures on Electrochemical Corrosion*, Plenum Press, New York, 1973.

14. D. A. Jones, *Principles and Prevention of Corrosion*, 2nd ed., Prentice Hall, Upper Saddle River, NJ, 1996.

15. M. Stern and A. L. Geary, Electrochemical polarization: I. A theoretical analysis of the shape of polarisation curves, *Journal of the Electrochemical Society*, 104 (1957), 56.

16. F. Mansfeld, The polarisation resistance technique for measuring corrosion currents, in *Advances in Corrosion Science and Technology*, Vol. 6, M. G. Fontana and R. W. Staehle, Eds., Plenum Press, New York, 1976, p. 163.

17. EG&G Princeton Applied Research, Model 352/252 SoftCorr™ II corrosion measurement and analysis software: user's guide, EG&G Instruments Corporation, Princeton, 1993.

18. W. J. Lorenz and F. Mansfeld, Determination of corrosion rates by electrochemical dc and ac methods, *Corrosion Science*, 21 (1981), 647.

19. F. Mansfeld, Don't be afraid of electrochemical techniques—but use them with care! *Corrosion*, 44 (1988), 856.

20. Y. J. Tan, B. Kinsella, and S. Bailey, An experimental comparison of corrosion rate measurement techniques: weight-loss measurement, linear polarisation, electrochemical impedance spectroscopy and electrochemical noise analysis, *Proceedings of Corrosion and Prevention '95*, Australasian Corrosion Association, Perth, WA, Australia, 1995.

21. I. Epelboin, M. Keddam, and H. Takenouti, Use of impedance measurements for the determination of the instant rate of metal corrosion, *Journal of Applied Electrochemistry*, 1 (1972), 71.

22. F. Mansfeld and W. J. Lorenz, Electrochemical impedance spectroscopy: application in corrosion science and technology, in *Techniques for Characterisation of Electrodes and Electrochemical Processes*, R. Varma and J. R. Selman, Eds., Wiley, New York, 1991.

23. D. A. Eden, K. Hladky, D. G. John, and J. L. Dawson, Electrochemical noise resistance, presented at Corrosion '86, Paper 274, NACE, Houston, TX, 1986.

24. R. A. Cottis, S. Turgoose, and R. Newman, *Corrosion Testing Made Easy: Electrochemical Impedance and Noise*, NACE International, Houston, TX, 1999.

25. F. Mansfeld and H. Xiao, Electrochemical noise analysis of iron exposed to NaCl solutions of different corrosivity, *Journal of the Electrochemical Society*, 140 (1993), 2205.

26. U. Bertocci, C. Gobrielli, F. Huet, and M. Keddam, Noise resistance applied to corrosion measurements: I. Theoretical analysis, *Journal of the Electrochemical Society*, 144 (1997), 31–37.

27. R. A. Cottis, The interpretation of electrochemical noise data, *Corrosion*, 27 (2001), 265.

28. Y. J. Tan, S. Bailey, and B. Kinsella, The monitoring of the formation and destruction of corrosion inhibitor films using electrochemical noise analysis, *Corrosion Science*, 38 (1996), 1681.

29. D. D. Macdonald, Some advantages and pitfalls of electrochemical impedance spectroscopy, *Corrosion*, 46 (1990), 229.

30. M. E. Orazem, P. Agarwal, and L. H. Garcia-Rubio, Critical issues associated with interpretation of impedance spectra, *Journal of Electroanalytical Chemistry*, 378 (1994), 51.

31. D. D. Macdonald, Application of electrochemical impedance spectroscopy in electrochemistry and corrosion science, in *Techniques for Characterisation of Electrodes and Electrochemical Processes*, R. Varma and J. R. Selman, Eds., Wiley, New York, 1991.

32. D. C. Silverman and J. E. Carrico, Electrochemical impedance technique: a practical tool for corrosion prediction, *Corrosion*, 44 (1988), 280.

33. D. C. Silverman, Rapid corrosion screening in poorly defined systems by electrochemical impedance technique, *Corrosion*, 46 (1990), 589.

34. C. H. Tsai and F. Mansfeld, Determination of coating deterioration with EIS: Part II. Development of a method for field testing of protective coatings, *Corrosion*, 49 (1993), 727.

35. G. W. Walter, A review of impedance plot methods used for corrosion performance analysis of painted metals, *Corrosion Science*, 26 (1986), 681.

36. F. Mansfeld and C. H. Tsai, Determination of coating deterioration with EIS: 1. Basic relations, *Corrosion*, 47 (1991), 958.

37. B. A. Boukamp, *Equivalent Circuit (EQUIVCRT.PAS)*, Version 3.96, 2nd ed., University of Twente, The Netherlands, 1989.

38. B. Kinsella, Y. J. Tan, and S. Bailey, Studies of CO_2 corrosion product scales using electrochemical impedance spectroscopy and surface characterisation techniques, *Corrosion*, 54 (1998), 835–842.

39. R. H. Hausler, The mechanism of CO_2 corrosion of steel in hot, deep gas wells, in *Advances in CO_2 Corrosion*, R. H. Hausler and H. P. Godard, Eds., NACE, Houston, TX, 1984, p. 72.

40. K. E. Heusler, The influence of electrolyte composition on the formation and dissolution of passivating films, *Corrosion Science*, 29 (1989), 131.

41. Y. J. Tan, S. Bailey, and B. Kinsella, Investigations on the formation and destruction processes of corrosion inhibitor films using electrochemical impedance spectroscopy, *Corrosion Science*, 38 (1996), 1545–1561.

42. Y. J. Tan, S. Bailey, and B. Kinsella, Studying the formation process of chromate conversion coatings on aluminium using continuous electrochemical noise resistance measurements, *Corrosion Science*, 44 (2002), 1277–1286.

43. Y. J. Tan, Monitoring localized corrosion processes and estimating localized corrosion rates using a wire-beam electrode, *Corrosion*, 54 (1998), 403.

44. Y. J. Tan, Wire beam electrode: a new tool for studying localised corrosion and other heterogeneous electrochemical processes, *Corrosion Science*, 41 (1999), 229.

45. C.-Y. Lee, Y. J. Tan, and A. M. Bond, Identification of surface heterogeneity effects in cyclic voltammograms derived from analysis of an individually addressable gold array electrode, *Analytical Chemistry*, 80 (2008), 3873.

46. A. J. Bard, H. D. Abruna, C. E. Chidsey, L. R. Faulkner, S. W. Feldberg, K. Itaya, M. Majda, O. Melroy, and R. W. Murray, The electrode/electrolyte interface: a status report, *Journal of Physical Chemistry*, 97 (1993), 7147–7173.

47. Y. J. Tan, Sensing electrode inhomogeneity and electrochemical heterogeneity using an electrochemically integrated multielectrode array, *Journal of the Electrochemical Society*, 156 (2009), C195–C208.

2

Probing Electrode Inhomogeneity, Electrochemical Heterogeneity, and Localized Corrosion

The uniform electrochemical corrosion models described in Chapter 1 suggest that uniform corrosion occurs when microscopic local anodic and cathodic sites are sufficiently small and are distributed randomly over an electrode surface. Uniform corrosion is the least dangerous form of corrosion because it occurs at predictable rates and can be controlled effectively by corrosion prevention technologies such as corrosion inhibitors and protective coatings. In practice, localized corrosion such as pitting and crevice corrosion is the most destructive form of corrosion and remains the most difficult, yet fundamentally important issue in corrosion science and engineering. Localized corrosion frequently causes unexpected premature failure of industrial and civil structures and is thus a major economic, safety, and reliability concern in many industries, such as oil and gas production, petrochemical and chemical processing, and aircraft and building maintenance. Major engineering materials such as stainless steels and nickel and aluminum alloys are often unsuitable for chloride-containing environments, due to susceptibility to localized corrosion.

Heterogeneous Electrode Processes and Localized Corrosion, First Edition. Yongjun Tan.
© 2013 John Wiley & Sons, Inc. Published 2013 by John Wiley & Sons, Inc.

Under certain environmental and material conditions, localized corrosion can be initiated by the formation of macroscopic local anodes and cathodes over an electrode surface. The fundamental reason for the formation of macroscopic local anodes and cathodes is that in electrochemical reactions the reactants do not need to be near each other spatially. As normal chemical reactions do, they collide with spatially separated anodes and cathodes. This characteristic permits distinctive separation of electrochemical reactions, leading to a nonuniform electrode–solution interface and localization of electrode surface chemistry and physics (i.e., electrochemical heterogeneity) [1–3].

Electrochemical heterogeneity is a ubiquitous phenomenon that has been recognized to play critical roles in many challenging issues in electrochemical science and engineering, such as localized corrosion, uneven electrodeposition, and nonuniform electrodissolution. The most common and probably the most important heterogeneous electrochemical process is localized corrosion, such as pitting and crevice corrosion. When localized corrosion occurs, there is a distinct separation of anodes and cathodes on the metal surface, and different electrochemical reactions occur in the anodic and cathodic areas. Electrons travel continuously from the anode to cathode areas through the electrode body, and at the same time, ions travel between the anode and cathode areas through the electrolyte, resulting in the most destructive form of corrosion, localized corrosion. Figure 2.1 illustrates localized corrosion due to the existence of localized anodes and cathodes distributing nonuniformly over an electrode surface.

Another important heterogeneous electrochemical process is cathodic protection of metals. In the case of cathodic protection of a metal structure using a sacrificial anode, protection current (galvanic current) is often not distributed uniformly over the metal structure surface. For example, locations that are far away from the sacrificial anode site often have lower protection current densities and thus may not be protected effectively. This is a major problem that has to be addressed when a cathodic protection system is designed. Other forms of heterogeneous electrochemical processes include nonuniform electrodeposition, such as the electroplating of a complexly shaped workpiece (a practical working electrode). Electrochemical reaction rates, which are determined by the density of applied electroplating

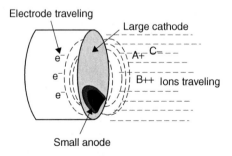

Figure 2.1 Localized corrosion due to electrochemical heterogeneity.

currents, can vary significantly over the surface of the workpiece. This can result in an electroplated layer of nonuniform thickness. Similar situations exist in almost all electrochemical industrial processes, such as electrotyping, electrometallurgy, electrowinning, and electromachining.

Unfortunately, electrochemical heterogeneity has not been confronted effectively in electrochemistry. It is evident that classical theories describing Faradaic electron transfer processes in dynamic electrochemistry are based on the assumption that the electrode surface is homogeneous [3,4]. Initial experimental studies used to validate electrochemical theories commonly employ mercury drop electrodes, where a truly homogeneous surface is indeed likely to be obtained. However, solid electrodes that are widely used today usually have an inhomogeneous surface where electrochemical heterogeneity can be initiated and develop. Studies by Compton and Banks suggest that electrochemists may need to recognize more widely the influence of surface inhomogeneity as a factor that introduces nonideal behaviors relative to those predicted on the basis of a homogeneous surface [5].

In this chapter we provide an overview of electrode inhomogeneity and electrochemical heterogeneity and their effects on the initiation of localized corrosion. Localized corrosion principles are reviewed; however, detailed descriptions of various forms of localized corrosion and their theories are not attempted because these have already been discussed by many authors, among them Tomashov [6], Evans [7], and Fontana [8]. Particular focus is on techniques such as scanning probes that could be applied for probing electrode inhomogeneity, electrochemical heterogeneity, and localized corrosion. Limitations in conventional techniques for measuring localized corrosion processes and mechanisms are discussed, leading to the concept of an electrochemically integrated multielectrode array: the wire beam electrode method.

2.1 PROBING ELECTRODE INHOMOGENEITY

Electrode inhomogeneity is a natural phenomenon that is observable on almost all solid electrodes that are in practical use today. Over the past two decades, significant research has been carried out to understand electrode inhomogeneity and its effects on the corrosion behavior of steels, stainless steels, and aluminum and magnesium alloys, due primarily to industry interest in improving the localized corrosion resistance of these important engineering materials.

There are many forms and sizes of electrode inhomogeneities from various origins. These include inhomogeneities preexisting on electrode surfaces, such as surface roughness and scratches, impurities, mill scales, surface flaws, metallurgical defects, precipitated phases, grain boundaries, dislocation arrays, and localized stresses. Electrode inhomogeneities can also develop over an electrode surface due to irregular chemical adsorption, selective dissolution, and damage in passive films. For example, under some environmental conditions, such as stresses in metals, passive films are sensitive to localized damage, causing major microscopic heterogeneity and localized corrosion [9]. It has been reported that the passive film

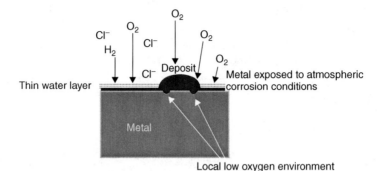

Figure 2.2 Electrode inhomogeneity due to environmental nonuniformity.

on passive metals has an extraordinarily large defect concentration, perhaps 1019 to 1021 per cubic centimeter [10]. The problem of electrode inhomogeneity is likely to be even more acute when an electrode surface is chemically or biologically modified or is covered by a monolayer [11].

Electrode inhomogeneity can also arise from microscopic or macroscopic nonuniformities in the environment due to differential aeration, concentration variation, local temperature, external stress, turbulent flow, erosion, and so on. For example, dusts or corrosion products deposited on a metal surface can create variations in the concentration of ions and oxygen over a metal surface. Figure 2.2 illustrates an inhomogeneous electrode surface exposed to a nonuniform atmospheric environment. The concentration of chemical species over the electrode surface could keep changing if there are differences in chemical reaction rates and mass transformation rates.

Challenges in studying electrode inhomogeneity are frequently due to technological difficulties in detecting and measuring inhomogeneities over electrode surfaces [3]. The most common and conventional means of detecting electrode inhomogeneities over an electrode surface are visual inspection, and observations using optical microscopy, scanning electron microscopy (SEM), and scanning tunneling electron microscopy (STM). Visual inspection of a metal surface can determine the sizes and shapes of major macro-sized inhomogeneities such as surface scraches, while a more detailed examination of a corroded surface can be carried out through optical microscopy and SEM. In particular, STM is capable of determining inhomogeneities in the nano dimensions and capturing surface changes in the millisecond time domain, while more powerful synchrotron light sources may well be able to probe structural changes occurring on the microsecond and submicrosecond time scales [12].

For example, STM has been used to probe inhomogeneities in metallurgical structures such as defects, precipitated phases, grain boundaries, and dislocation arrays. Brown et al. used STM to examine the nature, origin, and distribution of aluminum surface flaws [13]. It was established that even in 99.9% high-purity aluminum, the trace amounts of impurities were not distributed uniformly throughout the aluminum matrix. Such impurities were segregated in a "cellular honeycomb" structure enclosing fine volumes of relatively high-purity aluminum. The cellular boundaries, together with grain boundaries, form flaws in the protective oxide layer covering the macroscopic aluminum surface. Surface flaws were found to serve as cathodic sites during the corrosion of high-purity aluminum [13], leading to localized corrosion along grain boundaries [14].

Many surface analytical techniques and their combinations have been used to characterize chemical inhomogeneity over electrode surfaces. For example, x-ray radiography techniques such as x-ray photoelectron spectroscopy (XPS) and Auger electron spectroscopy (AES) have also been used to obtain information on chemical inhomogeneity over corrosion sample surfaces with minimum radiation damage to the sample. X-ray absorption and fluorescence measurements have been employed for in situ study of chemical and physical changes during electrochemical and corrosion processes [15]. Ryan et al. characterized the material chemistry of a stainless steel as a function of the proximity to sulfide inclusion using focused ion-beam/secondary-ion mass spectroscopy and found that a depletion of chromium during the processing of steel triggered a high-rate electrochemical dissolution at the edge of the sulfide inclusion [16]. These results were challenged by Schmuki et al. and considered un-verified by Auger microscopy and STM measurements [17].

During the past two decades, with the advent and refinement of such advanced scanning probes as the scanning Kelvin probe (SKP), scanning Kelvin probe force microscopy (SKPFM), the scanning reference electrode technique (SRET), the scanning vibrating electrode technique (SVET), local electrochemical impedance spectroscopy (LEIS), and scanning electrochemical microscopy (SECM), there has been a marked increase in research aimed at understanding electrode inhomogeneity and its effects on localized electrode processes. Many innovative experiments have been carried out and reported in the literature.

SKP and SKPFM are probe techniques that permit mapping of topography and Volta potential distribution on electrode surfaces. SKP and SKPFM scan the electric potential just above the electrolyte over an electrode surface to detect Volta potential differences over different parts of the electrode. SKPFM combines the SKP with atomic force microscopy (AFM) and uses much smaller probes and operates at much smaller distances from the surface, and thus SKPFM has an improved lateral resolution of better than 0.1 μm, compared to the classical SKP of 100 μm. The SKPFM technique is able to provide both Volta potential and topographical data with submicrometer resolution. In a typical application, Schmutz and Frankel [18] studied the surface inhomogeneity and localized corrosion of aluminum alloy AA2024-T3 using SKPFM. Large differences in the Volta potential of intermetallic particles and the matrix phase resulted in a potential map with high contrast that clearly identified the locations of the particles. All intermetallic particles, including the Mg-containing S-phase particles, had a Volta potential noble to that of the matrix. Surface films on the particles and the matrix were found to have strong effects on the Volta potential. A linear relationship was found between the Volta potential measured in air and the corrosion potential measured in aqueous solution, indicating that the Volta potential could be used to determine the practical nobility of the surface [18]. This relationship forms a basis for application of the Volta potential in understanding corrosion processes.

In another typical work, de Wit [19] employed a combination of techniques, including SKPFM, conventional SEM, and energy-dispersive x-ray spectroscopy (EDS) for probing a substantial increase in surface inhomogeneity in aluminum alloys with the addition of alloying elements in metals. SKPFM was used to measure intermetallic-related Volta potentials, while SEM and EDS were used to identify intermetallics and their compositions. It was found that inhomogeneity increases were due to the existence of complex metallurgical structures such as intermetallic phases, of size between 1 and 20 μm. The Volta potential differences between the major intermetallics Al_7Cu_2Fe, $(Al,Cu)_6(Fe,Cu,Mn)$, and Mg_2Si and the matrix were found to be about 300, 250, and -170 mV, respectively. Al_7Cu_2Fe and $(Al,Cu)_6(Fe,Cu,Mn)$ intermetallics behaved as cathodes, providing strong galvanic coupling with the matrix, while the Mg_2Si intermetallic behaved as an anode. This work confirmed that intermetallic particles play an extremely important role in the initiation process of localized corrosion attacks on passive surfaces.

Davoodi et al. [14] used SEM and EDS, in conjunction with SKPFM, AFM, and SECM, to characterize the microstructure and composition of intermetallics, including $Al_{12}(Mn, Fe)_3Si_2$, in an Al–Mn–Si–Zr alloy. Large Volta potential differences relative to the matrix were detected using SKPFM, while integrated AFM and SECM measurements detected extensive "tunnel-like" localized dissolution and deposition of corrosion products on the alloy surface. Integrated AFM topography and SECM electrochemical activity mapping with micro- or submicrometer resolution was considered to provide detailed information on dynamic localized corrosion processes of aluminum alloys, although this method was not able to clarify the relationship between the microstructure and corrosion initiation sites [20].

Jönsson et al. [20] confirmed that SKPFM provides information on the local nobility of the various intermetallic particles and phases on the submicrometer scale. The microstructure of commercial magnesium–aluminum alloy AZ91D and its effects on corrosion behavior have been determined using SKPFM and SKP. Volta potentials measured with SKP on both the η-Al_8Mn_5 and the β-$Mg_{17}Al_{12}$ phases in AZ91D showed more noble potentials than those of the α-magnesium phase. Aluminum-rich coring along the grain boundaries was found to result in measurable changes in the Volta potential. Ben-Haroush et al. [21] also found that the corrosion behavior of magnesium–aluminum alloys depended considerably on microstructure using various types of microscopic analysis, including optical microscopic, SEM, STM, and SKPFM techniques. Apachitei et al. [22] studied a Mg–Al–Ca-based alloy using SEM, XRD, and SKPFM and reconfirmed that intermetallic phases play an important role in corrosion processes. The SEM/EDS results revealed the two main types of intermetallics: Al–Ca(Sn,Sr) particles, which dominated the microstructure, and Al–Mn–Fe particles, fragmented and distributed along the extrusion direction. SKPFM revealed large Volta potential differences between intermetallics Al–Ca(Sn,Sr) (62 ± 7 mV) and Al–Mn–Fe (262 ± 18 mV) with an α-Mg matrix.

Research work, including that described above, has resulted in much better understanding of surface inhomogeneity and its effects on corrosion behavior, leading to significant improvements in the corrosion resistance of aluminum and magnesium alloys. One means of improving the corrosion resistance of alloys has been by reducing the contents of impurities such as Fe, Cu, and Ni in aluminum and magnesium alloys. It is often believed that electrode inhomogeneity can be alleviated by careful surface preparation and treatment. This is often true; however, in some cases surface treatment has also been found to be a source of surface inhomogeneity. Dasilva et al. [23] used STM to examine the effect of mechanical polishing on an aluminum surface. It was found that the preparation of ideally homogeneous surfaces could not be achieved readily by mechanical polishing, even when great care was taken in the procedures involved. Although a mirror-like surface finish can be generated with little visual evidence of significant heterogeneities, in reality, polishing powder was found to be embedded in the polished surface, giving rise to a newly introduced major surface inhomogeneity. Tanem et al. [24] demonstrated the crucial importance of sample treatment using a combination of SEM, SKP, and SKPFM which allows simultaneous determination of local microstructure and mapping of topography and Volta potential distribution on passive aluminum alloy surfaces. Mechanical polishing of the surface was found to produce a deformed layer distorting local chemistry and microstructure. Chemical and electrochemical processes were found to distort the local chemical composition through preferential dissolution of the aluminum matrix, leaving the intermetallics on the surface [24]. Turley and Samuels [25] made a similar observation by STM examination of inhomogeneities on copper surfaces mechanically polished using diamond abrasives and silicon carbide papers. Plastic deformation, slab-shaped cells, and recrystallized grains were found to be present on polished copper surfaces. Raicheva [26] also investigated surface inhomogeneities after applying various types of surface treatments, including mechanical grinding and polishing, chemical etching, and electrochemical polishing. Mechanical treatment was found to destroy and deform the surface layer of the metal and to increase its defect concentration and hence the surface energy and inhomogeneity. Electrochemical polishing was found to dissolve areas of increased energy, resulting in a more homogeneous surface with fewer electrochemically active areas, and thus reduced corrosion rates. However, electropolishing of low-purity metals and alloys could lead to surface enrichment of impurities and differential rates of matrix and second-phase dissolution, leading to increased surface inhomogeneity [26].

Extensive research, including the work described above, suggests that surface inhomogeneity should be a natural existence present on all practical electrode surfaces, regardless of their shape and size. Although electrode inhomogeneity could originate from dynamic processes such as the eruption of a passive film due to localized surface stress, most forms of electrode inhomogeneity are preexisting and distributed naturally on electrode surfaces. The significance of electrode inhomogeneity as a critical factor affecting the electrode process and electrochemical measurement has been recognized increasingly [5,11]; for example, Lee et al. discovered that inhomogeneity over a chemically or biologically modified electrode surface could drastically influence even the wave shape of its cyclic voltammograms [11]. The fundamental way in which electrode inhomogeneity affects electrode processes plays an important role in initiating electrochemical heterogeneity.

2.2 PROBING ELECTROCHEMICAL HETEROGENEITY AND LOCALIZED CORROSION

It is logical to expect that electrode inhomogeneities are the weakest locations over an electrode surface, where local electrochemical activities such as localized corrosion could be initiated. Indeed, much experimental evidence indicates that electrode inhomogeneities such as intermetallic particles, second phases, or impurities behave as vulnerable or susceptible sites for the initiation of electrochemical heterogeneity, causing the formation of pitting precursors (see Figure 2.1) [2,3]. For example, in stainless steels, the composition, density, and size of the sulfide inclusions have been reported to have a critical effect on the probability of initiation of pitting corrosion [27]. Intermetallics in aluminum alloys were found to weaken the passive film and form key sites for pit nucleation [28]. The β-phase structures were found to be responsible for the poor corrosion resistance of aluminum–magnesium alloys [29].

It should be pointed out that electrochemical heterogeneity differs from electrode inhomogeneity. Unlike surface inhomogeneity, which often preexists on an electrode due to uneven polishing, material defects, or irregular chemical adsorption, electrochemical heterogeneity describes electrode processes where two or more reactions take place simultaneously and dynamically at different parts of a surface, leading to the localization of reactions. Electrochemical heterogeneity is often initiated from preexisting surface inhomogeneities; however, it could also be initiated from an ideally homogeneous surface, especially if this surface is exposed to a nonuniform environment.

The fact that electrode inhomogeneity affects pitting corrosion susceptibility significantly is believed to be due to its close relationship with thermodynamic heterogeneity over an electrode surface. It is well known that a metal with a less defective oxide film is less susceptible to pitting, evaluated by pitting potential [28]. It has also been found that the pitting potential decreases as the surface roughness increases [30], aiding the nucleation of metastable pitting [31,32]. The pitting potential test [33] is probably the only traditional electrochemical method that can be used to characterize the pitting susceptibility by determining the potential at which the anodic current increases rapidly (i.e., the pitting potential). The nobler the pitting potential, obtained at a fixed scan rate in this test, the less susceptible is the alloy to initiation of pitting corrosion. Pitting potential appears closely related to the Volta potential and electrode inhomogeneities.

The existence of significant electrode inhomogeneity was found to lead to a nonuniform Volta potential distribution over a metal surface [4,18,19], and thus the Volta potential has been used to identify electrode inhomogeneities and the locations of anodic and cathodic sites over an electrode surface. Volta potential differences between high and low Volta potential areas would create a strong driving force for initiating electrochemical heterogeneity and localized electrode activities; however, they do not necessarily result in a large galvanic current flow [4,18,19]. This suggests that electrode inhomogeneity in the passive film would influence pit initiation primarily, playing a secondary role in pit growth. Indeed, sulfide inclusions in stainless steels were found to control the initiation phase of pitting, while the propagation phase was found to be determined by the electrochemistry of the steel in an extreme local environment [27], suggesting that the propagation of pitting is controlled by electrochemical heterogeneity.

An extensive experiment performed by Davoodi et al. [14] utilized a combination of several scanning probe techniques for studying the initiation of electrochemical heterogeneity on two aluminum alloys. In situ concurrent SKPFM and SECM measurements were carried out with micrometer resolution at open-circuit potential after various periods of specimen exposure in corrosive solutions. SKPFM, AFM, and SECM imaging of topography and electrochemical current were found to provide detailed information on the initiation of pitting precursors. SKPFM images of an Al–Mn–Si–Zr alloy revealed local dissolution of the aluminum matrix and the formation of corrosion products around a number of high-Volta-potential micrometer-sized intermetallic particles, forming tunnel-like pits. On an EN AW-3003 alloy, SKPFM images showed a large number of corroding sites and extensive local dissolution of the matrix in the boundary region adjacent to the high-Volta-potential intermetallic particles, resulting in a higher general corrosion loss. The integrated AFM/SECM probing of the EN AW-3003 surface in 20 mM NaCl solution containing 2 mM KI as mediator revealed local anodic current and topography changes associated with precursors of pitting in the passive potential region, and also extensive local active dissolution above the breakdown potential. A high SECM scan speed was used to catch fast ongoing electrochemical activities on the surface (e.g., passivity breakdown). With a small anodic polarization of about 100 mV, local SECM anodic current image was observed on the smooth surface of AFM image. Since the local current was stable for at least a couple of minutes, these sites were considered as pitting precursors, probably related to local dissolution of the matrix adjacent to certain intermetallic particles [14].

Pitting is the most common and characteristic form of localized corrosion. Over the past several decades, pitting has been studied extensively, with many interesting phenomena related to the pitting process identified. For example, pitting is found to require a passive external surface that behaves as the cathode of a galvanic cell to drive a high anodic dissolution current to flow into the pit. A pit may propagate for a short time and then "die" (metastable pitting), or it may continue to propagate essentially indefinitely (stable pitting). The metastable pits are believed to be very small in size and grow and repassivate within a few seconds. The consecutive formation and repassivation of microsize pits could lead to the occurrence of current oscillations [28]. The pitting process has also been described as a random and stochastic event [28].

A well-known assumption used to explain pitting initiation is that local damage in a protective passive film, usually by Cl^-, leads to the growth of a very small pit nucleus and metastable pit. This assumption is adopted in most theoretical models based on the breakdown of presumably uniform passive films [34–36]. However, these models should be considered to be speculative because pitting initiation from passive film breakdown sites has not been verified experimentally. An issue with these models is that they do not sufficiently consider the effects of electrode inhomogeneities such as inclusions on pitting initiation, although experimental evidence clearly indicates the significance of inclusions such as sulfide in pitting initiation [16,37] and that high-purity metals and amorphous alloys have far better pitting corrosion resistance than that of heterogeneous alloys that contain significant electrode inhomogeneities, such as inclusions.

Under certain environmental conditions, local instability and defects (microscale electrochemical heterogeneity) would propagate, causing stable growth of localized corrosion such as pitting and crevice corrosion and localized material deterioration, such as localized breakdown of coatings (macroscale electrochemical heterogeneity) [1–3]. Pitting requires an extended initiation period before propagating at an increasing rate in an autocatalytic manner. When localized corrosion continues to grow, a distinct separation of the anodic and cathodic areas results in rapid penetration of the metal at anodic areas and a distinctively different local chemical environment. In the propagation phase, corrosion current density within a growing pit would be very high, which draws Cl^- anions into the pit by electromigration to maintain charge neutrality. Pitting growth requires restriction of electrolytes mixing with bulk solution in order to build up and maintain an aggressive local environment within the pit that is acidified by hydrolysis of the dissolving metal cations. Figure 2.3 illustrates a typical metal surface under pitting corrosion. Szklarska-Smialowska provided a comprehensive review and concluded that the most important unaddressed issues in metastable pit studies are the processes leading to the formation of metastable pits and factors influencing the transition from metastable to stable pits [28].

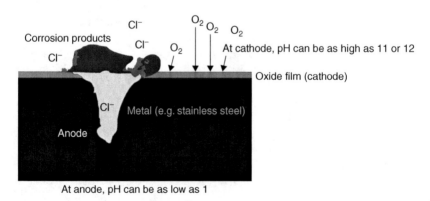

Figure 2.3 Electrode under pitting corrosion.

Electrochemical heterogeneity arising from environmental nonuniformities could also behave as a critical factor determining the initiation of localized corrosion attack. A typical example of localized corrosion due to environmental nonuniformity is corrosion of steel lampposts. As shown in Figure 2.4, serious localized corrosion of a lamppost occurs at an invisible location several centimeters below the soil surface. This is because this location has the most favorable corrosion conditions: differential aeration, a high level of moisture and conductivity, and the existence of a crevice between the lamppost and soil. Oxygen concentration cell effects and chemical changes in the crevice area lead to significant electrochemical heterogeneity and rapid crevice corrosion at the location.

Various forms of electrochemical heterogeneity are the root cause of different patterns of localized corrosion. Extensive research on diverse forms of localized corrosion has been carried out and reported over the past decades by leading scientists: among them, Tomashov [6], Evans [7], and Fontana [8]. Readers are referred to the classical literature for detailed descriptions and analysis of different forms of localized corrosion.

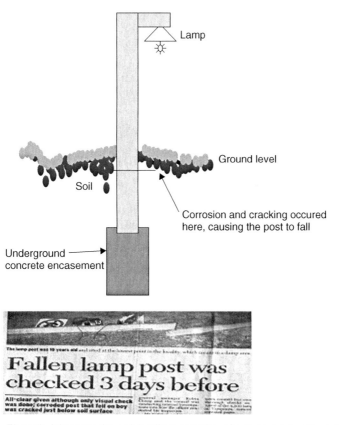

Figure 2.4   Electrode inhomogeneity and localized corrosion of a lamppost due to environmental nonuniformity.

2.3 OVERVIEW OF VARIOUS TECHNIQUES FOR PROBING LOCALIZED CORROSION

Detecting and measuring electrochemical heterogeneity and localized corrosion are challenging issues in modern electrochemistry. Conventional electrochemical methods using macrodisk electrodes have limitations in probing electrochemical heterogeneity and localized corrosion. These limitations can be illustrated by examining traditional methods of electrode potential measurement. Conventional potential measurement using a macrodisk electrode, a reference electrode, and a voltmeter is only applicable to measuring the potential of uniform electrodes where the potential at any location equals that of the entire electrode surface. If an electrode is heterogeneous, this method only detects a potential that is a mixture of contributions from many local potentials, none of which we can evaluate independently. When a one-piece electrode is used, it is impossible to measure the galvanic current that flows between localized anodic and cathodic sites in the electrode body since an ammeter cannot be inserted between anodic and cathodic sites which are located on a single piece of metal surface. It is also known that traditional electrochemical theories are based on a one-piece electrode with an ideally uniform working surface. The fundamental equations and theories describing the electrochemical thermodynamics and kinetics in Chapter 1, the Nernst and Butler–Volmer equations and the Wagner–Traud mixed potential theory, are all based on a homogeneous electrode surface and a uniform electrochemical mechanism. Traditional electrochemical techniques based on these equations, such as the Tafel polarization technique, the linear polarization technique, and ac impedance spectroscopy, in principle, are applicable only to homogeneous electrode surfaces for measuring uniform corrosion rates. It is evident that there is a major technological gap between conventional electrochemistry over uniform surfaces, and heterogeneous electrochemistry over nonuniform surfaces.

To characterize localized corrosion and other heterogeneous electrochemical processes, parameters related to local electrochemical processes, such as local electrochemical potential and local electrochemical kinetics, have to be determined. However, conventional electrochemical techniques have major limitations in measuring local electrochemical parameters and in determining local electrochemical kinetics. For example, electrochemical techniques such as linear polarization and ac impedance spectroscopy do not measure the rate of localized corrosion at a specific location. Traditional visual, optical, and SEM observations of test specimens for determining localized corrosion are slow and require periodic removal of test specimens from the corroding environment, which is cumbersome and may alter the progress of heterogeneous electrochemical processes.

In particular, research on the early stages of localized corrosion is difficult because it requires in situ methods which enable electrochemical heterogeneity-related parameters to be measured when local electrode processes are initiating, propagating, or terminating dynamically. Parameters that could be used to evaluate electrochemical heterogeneity include local chemical compositions and concentrations, local electrode potentials, galvanic currents flowing in electrolytic and metallic phases, and local electrochemical impedances. Since these local chemical and electrochemical parameters are difficult to measure by conventional analytical or electrochemical means, a major effort has been focused on the development of new techniques, such as scanning probes, microelectrodes, and electrochemical noise analysis techniques.

Techniques developed in a rich variety of subdisciplines have necessarily been employed in electrochemical and corrosion research to address problems relating to the structure, material properties, and dynamics of the electrode–solution interfacial region and its components [12]. During the past two decades, the advent of advanced physical and electrochemical techniques, in particular scanning probe techniques, including AFM, SKP, SKPFM, SRET, SVET, LEIS, and SECM, has facilitated substantial progress in localized corrosion and research as to its inhibition. Scanning probe techniques scan and detect local electrode potentials, galvanic currents, and local electrochemical impedances at the metal surface or metal–electrolyte interface for studying localized corrosion activities.

SRET and SVET are early scanning probe techniques designed to probe local ionic currents flowing in the electrolyte phase by detecting small potential variations over electrode surfaces where local electrode processes occur [38,39]. In SRET, this is usually achieved by scanning a passive reference probe parallel and in close proximity to the metal surface. A typical probe arrangement utilized in the SRET consists of two platinum tips whose diameters are on the order of 1 to 5 μm. These tips are housed within a single unit in close lateral proximity but spaced vertically from each other by a few millimeters. Figure 2.5 illustrates the SRET measurement principle.

SRET is an in situ technique for measuring microgalvanic current flowing close to the surface of an electrochemically heterogeneous specimen in an electrolyte. It is considered to be a tool for examining electrochemical heterogeneity due to surface contamination, surface roughness, stress defects, filiform corrosion, and paint and coating applications [38]. The technique relies on detecting extremely small potential variations, which exist on the surface of electrochemically nonuniform material surface exposed to an electrolyte. For example, when a stainless steel surface is under pitting corrosion, there exists a flow of metal ions in the solution above and around the pit site. This current movement leads to potential variations that SRET detects and quantifies. Resolution of the technique is dependent not only on the proximity of anodic and cathodic sites, but also on the magnitude of the corrosion current. Thus, if the flow of the current is large enough, the variation in potential could be detected. By scanning a passive reference probe with a fine capillary tip parallel and in close proximity to the metal surface, the potential distribution in the solution can be measured. The scanning probe arrangement utilized to measure these potential variations usually consists of two platinum tips on the order of 1 to 5 μm in diameter configured as differential inputs to the measuring electronics (Figure 2.5). In SRET measurements, the probe moves horizontally to build up two dimensionally resolved data. The distance between the work surface and the probe tip is kept constant during the experiment. As the probe scans over the metal

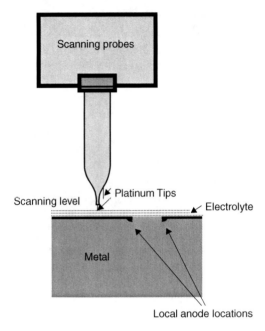

Figure 2.5   Probe tips and SRET measurement.

surface, there will be an *IR* drop between the two tips of the probe. Potential variations caused by ionic current flows within the electrolyte could be measured if the probe is within close proximity of corrosion sites and if the electrolyte conductivity is not too high. The electrochemical cell is controlled separately by the use of a potentiostat and reference electrode system. The SRET data can be displayed as linescans or two-dimensional area maps offering a spatially resolved representation of the electrochemical heterogeneity present at an electrode surface. Figure 2.6 shows a SRET map measured from a mild steel array surface after about 10 hours' immersion in an Evans solution, together with a photo showing the wire beam electrode (WBE) surface after 4 days' immersion [40].

SRET has been applied widely in the study of various forms of localized corrosion, such as pitting initiation, intergranular corrosion, surface film breakdown, stress corrosion cracking, and galvanic corrosion [38,41–45]. However, it should be noted that SRET measures currents only in the solution phase, not precisely at the metal–solution interface. Thus, the technique cannot detect ionic currents that flow at the metal–solution interface [39] and has limitations in the study of corrosion systems that generate only small corrosion currents or that have anodic and cathodic sites not well separated. The sensitivity of SRET is highly dependent on the corrosion current value, electrolyte conductivity, and probe proximity over corrosion sites. The movement of scanning tips could interfere with electrochemical processes occurring over an electrode surface, especially when a corrosion product forms over an electrode surface.

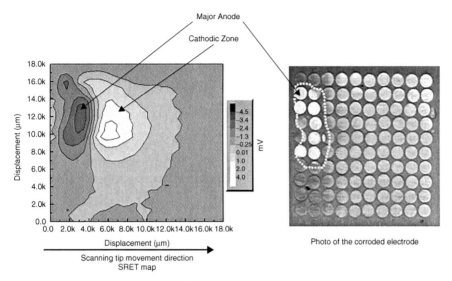

Figure 2.6   Typical SRET map together with a photograph showing a corroded surface. (From [40].)

SVET was developed to improve the sensitivity of detecting local electrochemical and corrosion events by operating with a nonintrusive scanning vibrating probe that measures the electric field generated by localized electrode processes over an electrode–solution interface [46]. The probe vibration is controlled by a piezoceramic displacement device that allows vibration amplitudes generally from 1 to 30 μm (perpendicular to the sample surface). It is an ac technique; thus, high system sensitivity is achieved via a differential electrometer in conjunction with an integrated lock-in amplifier. This technique has been used in many types of research and testing work, in particular for localized damage to organic coating films [46,47] and in inhibitor studies [48]. For example, Montemor et al. [48] investigated the effects of cerium nitrate and lanthanum nitrate on the pretreatment of AZ31 Mg alloy using SVET in conjunction with conventional potentiodynamic polarization and open-circuit potential measurements. They found that pretreatment reduced the corrosion activity of the alloy in chloride ion–containing solutions. SVET has shown better sensitivity; however, it still has limitations in detecting small ionic currents exactly at the metal–solution interface.

LEIS [49] is another scanning probe technique that can map the ac impedance distribution over an electrode surface. In LEIS a sinusoidal voltage perturbation between the working electrode and a reference electrode is maintained by driving an ac current between the working electrode and a distant counter electrode with a potentiostat, similar to what occurs in traditional ac impedance methods. Local ac impedances are then derived from the ratio of the applied ac voltage and the local ac solution current density. The local ac current density is obtained by measuring potential differences near the electrode surface using a probe consisting of two microelectrodes. By measuring the ac potential differences between the microelectrodes and knowing their separation distance and the solution conductivity, the local ac solution current density is derived. Impedance maps obtained from LEIS measurement could detect electrochemical heterogeneity and localized corrosion activities over an electrode surface, while traditional ac impedance of this electrode gave little indication of its presence [49]. LEIS has been applied in various coating [50] and corrosion inhibitor research [51]. For example, Barranco et al. [51] used LEIS to study the behavior of inhibitor cerium ions containing sol–gel coatings to protect against AZ91 magnesium alloy corrosion.

The scanning electrochemical microscope (SECM) [52,53] is a tool that enables us to perform the difficult tasks of detecting localized chemistry changes by means of variously designed scanning probes. SECM is a scanning electrochemical probe that detects amperometrically surface-generated electroactive ions or molecules in the solution phase as a function of spatial location with an electrochemically sensitive or ion-selective ultramicroelectrode tip. It has been used extensively for topography mapping, surface modification, and redox reactivity imaging [52,53]. When used to make high-resolution chemical concentration maps of corroding metal surfaces, it was found that minor spatial fluctuations on the picoampere scale in passive current density of type 304 stainless steel in dilute aqueous chloride solution were related directly to the subsequent initiation of pitting corrosion [54]. SECM measurements suggest that sulfide inclusions should be the initiation sites, although a definite coincidence could not be established [54]. SECM in ferrocyanide and bromide solutions was used to locate pitting precursors on polycrystalline Ti [32].

It should be noted that scanning probe techniques, including SRET, SVET, LEIS, and SECM, can detect ionic currents, carried by ions in the electrolyte phase, flowing over a corroding metal surface; however, they are unable to measure the currents flowing exactly at the metal–solution interface. For this reason, they may not be able to detect all ionic currents accurately, especially those flowing at the metal–solution interface. Scanning probe techniques commonly operate in a relatively specific and localized area, and thus, in many circumstances, the scan image does not necessarily represent the full details of an electrode process that involves different reactions occurring simultaneously over distinctively separated electrode areas. When investigating corrosion, it is difficult to ensure that a scanning tip is positioned correctly over a pit precursor unless the precursor is generated by the probe tip itself. This implies that successful imaging of a natural pit initiation by scanning probes could depend on "luck" in experiments.

Each technique has advantages and limitations, so different techniques are often combined and applied in a novel and synergistic manner. For example, traditional optical microscopy, SEM, and EDS are often used with scanning probe techniques to provide topographical and chemical information that is often critical for explaining electrochemical heterogeneity and localized corrosion activities. Analysis of electroctrochemically active sites can be carried out using SEM, EDS, AFM, and in situ confocal laser scanning microscopy. Using the scanning microreference electrode technique combined with SEM and EDS, Shao et al. [55] found that the micropitting on Al 2024–T3 began immediately following its exposure to

NaCl solution at open-circuit potential. Some corrosion-active nuclei were found to develop with time, whereas others disappeared. Most of the micropits were found to be associated with second-phase particles, especially S-phase particles (Al_2CuMg). Rough surface finishes [56] and nonmetallic inclusions in stainless steels [16,27] have also been reported to nucleate metastable pitting.

The pitting potential test [33] using cyclic polarization is probably the only standardized traditional electrochemical method that is considered capable of measuring relative localized corrosion susceptibility. As shown in Figure 2.7, this method involves anodic polarization of a specimen until localized corrosion is initiated, as indicated by a large increase in the current applied. An indication of the susceptibility to the initiation of localized corrosion in this test method is given by the potential at which the anodic current increases rapidly (i.e., the breakdown potential or the pitting potential). Conventional understanding is that the pitting potential E_{pit} is the potential above which pits are initiated, while the repassivation potential E_p obtained using a reverse scan is the potential below which pits repassivate. The nobler the pitting potential, obtained at a fixed scan rate in this test, the less susceptible is the alloy to initiation of localized corrosion. It is considered that if the corrosion potential of a system is at or above E_{pit}, new pits would initiate and existing pits would propagate. If the corrosion potential of a system is between E_{pit} and E_p, existing pits would propagate but no new pits would form. If the corrosion potential is below E_p, there would be no pitting. There have also been numerous attempts to use the amount of hysteresis in the cyclic scan as a measure of localized corrosion susceptibility, with varying degrees of success.

However, it has been found that at least in chloride solution, pits can nucleate and even grow at potentials below the pitting potential [30]. The pitting potential does not mark the boundary between pitting and no pitting, although it is believed

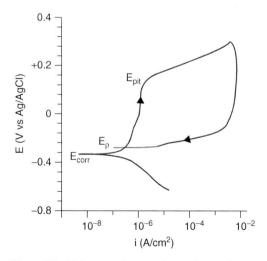

Figure 2.7 Pitting potential test using cyclic polarization.

to separate the potential above which nucleated pits can propagate indefinitely to achieve a state of stable growth and below which stable growth cannot be achieved [30]. In cyclic polarization measurements, large scatter in the breakdown potential and its dependence on scan rate are often experienced. These are thought to be due to the sensitivity of pit initiation to the initial conditions and to the time dependence of the localized corrosion-site chemistry. Experimentally induced artifacts such as inappropriate scan rate, large uncompensated solution resistance, and nonuniform current and potential distributions are common sources of errors. It should also be noted that the results of the cyclic polarization test are not intended to correlate in a quantitative manner with the rate of localized corrosion or the distribution of pits. Pitting potential appears closely related to electrode inhomogeneities since it has been found that the pitting potential decreases as the surface roughness increases [30].

Although scanning probes are able to detect growing pits, they may be unable to observe dynamic initiating pits. A method of detecting dynamic metastable pitting events often relies on the recording of current transients, or electrochemical noise, which is supposed to be due to the evolution of metastable corrosion pits. Electrochemical noise analysis (ENA) is a technique proposed and utilized for detecting localized corrosion and for measuring corrosion rates [57–66]. It has received considerable attention since Iverson's study [65] on electrode potential fluctuations and corrosion processes. Iverson suggested the possibility of studying the corrosion processes and quantitatively measuring the rate of corrosion by analyzing electrode voltage fluctuations [65].

There are two distinctly different approaches to measuring electrochemical noise: measuring the potential noise of a single sample relative to a low-noise reference electrode using a voltmeter and measuring current noise between a pair of identical electrodes using an ammeter. Figure 2.8a shows a device for detecting potential noise from a corroding electrode surface. Hladky and Dawson developed a device with a double reference electrode [57] (Figure 2.8b), and applied more sensitive detection instruments in order to detect characteristic potential noise patterns. Later, Hladky and Dawson modified the device by replacing the conventional reference electrode with an identical second working electrode (Figure 2.8c). The potential difference between the two identical working electrodes was sampled by a voltmeter as potential noise.

The prime attraction of ENA is probably its possibility of early detection of localized corrosion by detecting "noise signatures," proposed to detect localized corrosion by recognizing characteristic noise patterns (often referred to as noise signatures) in the time domain [57] or in the frequency domain [58]. In the time domain, Hladky and Dawson discovered that electrodes undergoing either pitting or crevice corrosion would generate quite distinct noise signatures in potential fluctuations [57]. They found that potential noise associated with pitting corrosion initiation is characterized by a series of sharp decreases in the electrode potential followed by exponential recoveries. They concluded that pitting or crevice corrosion attack can be detected within seconds of its initiation. Figure 2.9a illustrates typical potential noise patterns observable during pitting corrosion initiation.

Later, amplitude spectra of low-frequency electrochemical noise was found to correlate with the rate and pattern of corrosion attack [67]. It has been claimed that a typical $1/f$ noise spectrum with a roll-off slope of -10 or -20 dB/decade indicates pitting corrosion and that a spectrum with a roll-off slope of -40 dB/decade indicates general corrosion [67]. Figure 2.9b illustrates a typical potential noise amplitude spectrum observable during pitting corrosion initiation. Based on these findings, they advocated the use of characteristic noise patterns as indicators of localized corrosion [57,58,67] and suggested the possibility of a nonperturbative electrochemical corrosion monitoring technique capable of detecting localized

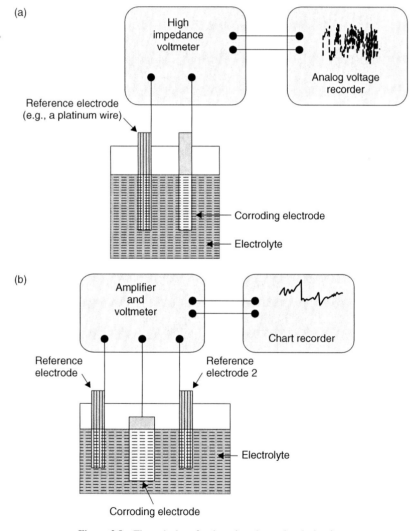

Figure 2.8 Three devices for detecting electrochemical noise.

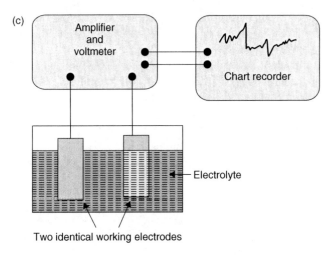

Figure 2.8 (*Continued*)

corrosion. Noise signatures have been recognized to be useful indicators of localized corrosion and tools to reveal fundamental information about the breakdown of passive film, and the incubation, propagation, and repassivation processes of localized corrosion.

Eden et al. [59] were probably the first to use a device (Figure 2.8c) to detect the current noise between two nominally identical electrodes using a zero-resistance ammeter while monitoring the potential noise of the coupled electrodes relative to a low-noise reference electrode (or to a third identical electrode). It has been appreciated that the combination of electrochemical potential and current noise is more powerful than individual measurements. Obvious advantages of this device include that it uses no applied external polarization signal and that it collects potential and current noise data essentially simultaneously. This device is now effectively the standard apparatus used to measure electrochemical noise. Based on this design concept, many corrosion sensors have been fabricated for various applications [66]. Having obtained potential and current noise time records, many methods can be used to analyze the data, as summarized in a comprehensive review by Cottis [63]. The identification of noise signatures may be undertaken by a variety of means; for example, examination of the time record trace may give an indication of the types of processes occurring. The initiation of pitting was characterized by sharp fluctuations of potential and current [57,63]. Typical potential and current fluctuation patterns of pitting initiation of carbon steel are shown in Figure 2.10. Frequency-domain analysis is believed to provide insight into the fundamental mechanisms that are operative and give information about low-frequency impedance of the interface [57,63]. The power spectra method is most common for noise analysis, although wavelet methods are also applicable. Two common ways to estimate a power spectrum are fast Fourier transform and the maximum entropy method [62,63].

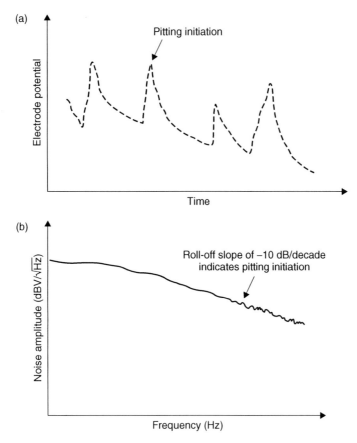

Figure 2.9 Typical potential noise from an electrode undergoing pitting corrosion.

Although localized corrosion sensing by means of ENA has progressed significantly over the past decades, there are still several unanswered concerns that limit its practical application. First, the origin and production mechanism of electrochemical noise have not yet been fully understood, although a number of noise-generating processes, such as metastable pitting, turbulent mass transport, particle impact, bubble nucleation, and separation, have been identified in the general scientific literature [63]. Since there is no suitable experimental technique that can be used to correlate noise signatures directly with localized corrosion activities occurring at a specific location of an electrode surface, many workers who have been studying the application of noise signatures to the identification of localized corrosion only tried to relate certain characteristic noise features they observed during certain periods of experiment to localized corrosion such as pits identified visually or microscopically after completion of the experiment. The expectation is that the number of "peaks" in potential fluctuation data for a certain immersion period could equal the number of pits counted by the optical microscope after immersion. However, it is

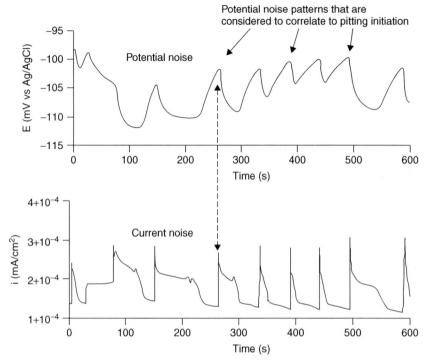

Figure 2.10   Electrochemical noise from a pitting corrosion system with a dual mild steel electrode exposed to a solution containing 1000 ppm NaNO$_2$ and 4000 ppm NaCl for 10 hours. (From [68].)

well known that, usually, only a small portion of the peaks observable in potential fluctuation data lead to stable corrosion pits, and thus this approach is valid only if there are means of identifying the "valid" peaks that lead to the formation of stable pits.

Another issue is the complexity of noise detection in practical corrosion systems. Although the noise measurement is reasonably straightforward, care is needed to avoid instrument noise and extraneous noise, aliasing, and quantization [62]. In real corrosion systems, both potential and current noise may be low in amplitude and difficult to measure. For example, the measurement of potential noise is expected to be particularly difficult for uniform corrosion of large electrodes because the power spectral density of potential noise is expected to be inversely proportional to the specimen area [62]. The sensitivity of conventional noise detection is often not high enough to recognize relatively small noise activities that are associated with the initiation of tiny localized corrosion sites. This is because the traditional noise detection method using a one-piece electrode (or two identical short-circuited electrodes) measures only a mixed or averaged potential and its fluctuations over the entire electrode surface. The initiation of pitting corrosion usually involves only a small electrode area; therefore, such an event could only result in a very small and often invisible fluctuation in an overall mixed or averaged electrode potential. Furthermore, asymmetry between the two current-measuring electrodes is considered to present a major practical and theoretical challenge [63]. In real corrosion systems, the two nominally identical electrodes are typically not similar, and the coefficient of variation is strongly dependent on the asymmetry between the two electrodes.

A further issue involves the difficulties associated with dc drift or trend removal. If the dc drift is not removed, it will significantly influence noise analysis. If the drift consists of a linear change in the mean divided by time, it can be removed simply by subtracting the linear regression line from the data, a common method of treating drift, especially prior to spectral estimation [63]. It is very difficult to remove complex forms of dc drift, although a method called *moving-average removal* has been used successfully in noise resistance analysis [64]. Another problem with dc drift is that its existence implies that the signal is nonstationary and thus that virtually all standard analysis procedures become invalid [63]. This raises an additional issue as to the reliability of some analysis methods; for example, the localization index is considered to be an unreliable localized corrosion indicator in some cases and thus must be used with care [63, 69].

Although some controversial issues still exist in the interpretation of electrochemical noise data, the noise signatures are considered to be valuable indicators of localized breakdown of passive film and the incubation, initiation, propagation, and repassivation processes of localized corrosion [63]. Electrochemical noise should be useful for identifying periods when the corrosion processes become unstable and to recognize when the probability of localized corrosion is high.

It should also be noted that direct experimental observation to correlate electrode current and potential transients to specific nucleation events or a single metastable pit event is, in fact, difficult. For this reason, results from noise analysis of pitting events are somewhat speculative. Many experiments have been carried out to observe, characterize, and understand metastable pits by potential transients. A typical experiment was carried out by Hashimoto et al. [70], who reported that metastable pitting events at open circuit resulted in potential transients with a typical shape consisting of a rapid decrease followed by a slow increase (see Figure 2.11a). However, in the same paper, Hashimoto et al. also showed distinctively different potential transients with the typical shape of a rapid shift toward a less negative direction, followed by a slow recovery (see Figure 2.11b). This suggests that pitting initiation detection is still an important, yet underdeveloped field that requires further research.

It should also be noted that although research using new techniques described above shed new light on the difficult issue of probing electrode inhomogeneity, electrochemical inhomogeneity, and localized corrosion, these techniques have also shown limitations. For example, scanning probe techniques such as SRET measure only the currents in solution, and not exactly at the electrode–electrolyte interface; the scanning results may not sufficiently reflect the complex distribution of ionic current on an electrode surface or in a pit. Scanning probes commonly operate

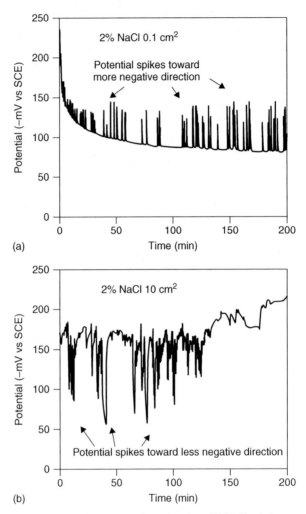

Figure 2.11 Corrosion potential fluctuations of pure iron in a 2% NaCl solution containing 200 ppm of NaNO₂ immediately after immersion. A stable and growing pit was observed after approximately 130 minutes. From [70].)

in a relatively specific area; thus, the scan images may not necessarily represent the full details of electrode processes that involve different reactions occurring simultaneously over distinctively separated electrode areas. As a result, basic questions in this field remain unanswered. For example, according to a current pitting corrosion model, pitting initiation occurs only at very few "breakdown" or "eruption" sites among a large number of vulnerable electrode inhomogeneity sites. This suggests that a large number of "preferential" electrode inhomogeneity sites remain inactive during the pitting process. Is this true? Is the pitting corrosion initiation truly a rare event occurring only at a few breakdown points of a passive film? What types of electrode inhomogeneity would lead to stable electrochemical heterogeneity leading to pitting corrosion? Conventional electrochemical techniques and models are insufficient to address these major issues.

REFERENCES

1. J. O'M. Bockris, R. E. White, B. E. Conway, Eds., *Modern Aspects of Electrochemistry*, Vol. 31, Springer-Verlag, New York, 1998.

2. Y. J. Tan, Sensing electrode inhomogeneity and electrochemical heterogeneity using an electrochemically integrated multielectrode array, *Journal of the Electrochemical Society*, 156 (2009), C195–C208.

3. Y. J. Tan, Understanding the effects of electrode inhomogeneity and electrochemical heterogeneity on pitting corrosion initiation on bare electrode surfaces, *Corrosion Science*, 53 (2011), 1845–1864.

4. A. J. Bard and L. R. Faulkner, *Electrochemical Methods: Fundamentals and Applications*, 2nd ed., Wiley, New York, 2000.

5. R. G. Compton and C. E. Banks, *Understanding Voltammetry*, World Scientific, London, 2007.

6. N. D. Tomashov, *The Theory of Corrosion and Protection of Metals*, Macmillan, New York, 1966.

7. U. R. Evans, *An Introduction to Metallic Corrosion*, 3rd ed., Edward Arnold, London, 1981.

8. G. Fontana, *Corrosion Engineering*, 3rd ed., McGraw-Hill, New York, 1987.

9. W. Li, Pitting corrosion induced by the interaction between microstructures of iron-based alloy in chloride-containing solution, *Journal of Materials Science Letters*, 21 (2002), 1195–1198.

10. H. J. Engll, Report on workshop group deliberations and discussions, *Corrosion Science*, 29 (1989), 119–127.

11. C.-Y. Lee, Y. J. Tan, and A. M. Bond, Identification of surface heterogeneity effects in cyclic voltammograms derived from analysis of an individually addressable gold array electrode, *Analytical Chemistry*, 80 (2008), 3873–3881.

12. A. J. Bard, H. D. Abruna, C. E. Chidsey, L. R. Faulkner, S. W. Feldberg, K. Itaya, M. Majda, O. Melroy, and R. W. Murray, The electrode/electrolyte interface: a status report, *Journal of Physical Chemistry*, 97 (1993), 7147–7173.

13. G. M. Brown, K. Shimizu, K. Kobayashi, G. E. Thompson, and G. C. Wood, Further evidence for the presence of residual flaws in a thin oxide layer covering high purity aluminum, *Corrosion Science*, 34 (1993), 2099–2104.

14. A. Davoodi, J. Pan, C. Leygraf, and S. Norgren, Integrated AFM and SECM for in situ studies of pitting corrosion of Al alloys, *Electrochimica Acta*, 52 (2007), 7697–7705.

15. H. S. Isaacs, J. H. Cho, M. L. Rivers, and S. R. Sutton, In situ x-ray microprobe study of salt layers during anodic dissolution of stainless steel in chloride solution, *Journal of the Electrochemical Society*, 142 (1995), 1111–1117.

16. M. P. Ryan, D. E. Williams, R. J. Chater, B. M. Hutton, and D. S. McPhail, Why stainless steel corrodes, *Nature*, 415 (2002), 770–774.

17. P. Schmuki, H. Hildebrand, A. Friedrich, and S. Virtanen, The composition of the boundary region of MnS inclusions in stainless steel and its relevance in triggering pitting corrosion, *Corrosion Science*, 47 (2005), 1239–1250.

18. P. Schmutz and G. S. Frankel, Characterization of AA2024-T3 by scanning Kelvin probe force microscopy, *Journal of the Electrochemical Society*, 145 (1998), 2285–2295.

19. J. H. W. de Wit, Local potential measurements with the SKPFM on aluminum alloys, *Electrochimica Acta*, 49 (2004), 2841–2850.

20. M. Jönsson, D. Thierry, and N. LeBozec, The influence of microstructure on the corrosion behavior of AZ91D studied by scanning Kelvin probe force microscopy and scanning Kelvin probe, *Corrosion Science*, 48 (2006), 1193–1208.

21. M. Ben-Haroush, G. Ben-Hamu, D. Eliezer, and L. Wagner, The relation between microstructure and corrosion behavior of AZ80Mg alloy following different extrusion temperatures, *Corrosion Science*, 50 (2008), 1766–1778.

22. I. Apachitei, L. E. Fratila-Apachitei, and J. Duszczyk, Microgalvanic activity of an Mg–Al–Ca-based alloy studied by scanning Kelvin probe force microscopy, *Scripta Materialia*, 57 (2007), 1012–1015.

23. M. F. Dasilva, K. Shimizu, K. Kobayashi, P. Skeldon, G. E. Thompson, and G. C. Wood, On the nature of the mechanically polished aluminum surface, *Corrosion Science*, 37 (1995), 1511–1514.

24. B. S. Tanem, G. Svenningsen, and J. Mardalen, Relations between sample preparation and SKPFM Volta potential maps on an EN AW-6005 aluminum alloy, *Corrosion Science*, 47 (2005), 1506–1519.

25. M. Turley and L. E. Samuels, The nature of mechanically polished surfaces of copper, *Metallography*, 14 (1981), 275–294.

26. S. N. Raicheva, The effect of the surface state on the electrochemical behavior of copper electrodes, *Electrochimica Acta*, 29 (1984), 1067–1073.

27. D. E. Williams, M. R. Kilburn, J. Cliff, and G. I. N. Waterhouse, Composition changes around sulphide inclusions in stainless steels, and implications for the initiation of pitting corrosion, *Corrosion Science*, 52 (2010), 3702–3716.

28. Z. Szklarska-Smialowska, Pitting corrosion of aluminum, *Corrosion Science*, 41 (1999), 1743–1767.

29. A. Pardo, M. C. Merino, A. E. Coy, F. Viejo, R. Arrabal, and S. Feliú, Jr., Influence of microstructure and composition on the corrosion behavior of Mg/Al alloys in chloride media, *Electrochimica Acta*, 53 (2008), 7890–7902.

30. G. T. Burstein, C. Liu, R. M. Souto, and S. P. Vines, The origins of pitting corrosion, *Corrosion Engineering, Science, and Technology*, 39 (2004), 25–30.

31. Y. Zuo, H. Wang, J. Zhao, and J. Xiong, The effects of some anions on metastable pitting of 316L stainless steel, *Corrosion Science*, 44 (2002), 13–24.

32. L. F. Garfias-Mesias, M. Alodon, P. I. James, and W. H. Smyrl, Determination of precursor sites for pitting corrosion of polycrystalline titanium by using different techniques, *Journal of the Electrochemical Society*, 145 (1998), 2005–2010.

33. ASTM G61-86, *Standard Test Method for Conducting Cyclic Potentiodynamic Polarization Measurements for Pitting Corrosion Susceptibility of Iron-, Nickel-, or Cobalt-Based Alloys*, ASTM, West Conshohocken, PA, 2009.

34. J. A. Richardson and G. C. Wood, Study of the pitting corrosion of Al by scanning electron microscopy, *Corrosion Science*, 10 (1970), 313–323.

35. T. P. Hoar, D. C. Mears, and G. P. Rothwell, The relationships between anodic passivity, brightening and pitting, *Corrosion Science*, 5 (1981), 279–289.

36. C. Y. Chao, L. F. Lin, and D. D. Macdonald, A point defect model for anodic passive films, *Journal of the Electrochemical Society*, 128 (1981), 1187–1194.

37. J. Stewart and D. E. Williams, The initiation of pitting corrosion on austenitic stainless steels on the role and importance of sulphide inclusions, *Corrosion Science*, 33 (1992), 457–474.

38. H. S. Isaacs and B. Vyas, Scanning reference electrode technique in pitting corrosion, in *Electrochemical Corrosion Testing*, ASTM STP 727, ASTM, West Conshohocken, PA, 1981.

39. H. S. Isaacs, M. P. Ryan, and L. J. Oblonsky, *Proceedings of the Research Topic Symposium, NACE Corrosion '97*, NACE, Houston, TX, 1997, p. 65.

40. Y. J. Tan, N. N. Aung, T. Liu, et al., Novel corrosion experiments using the wire beam electrode (I–IV), *Corrosion Science*, 48 (2006), 23–78.

41. R. Akid and D. J. Mills, A comparison between conventional macroscopic and novel microscopic scanning electrochemical methods to evaluate galvanic corrosion, *Corrosion Science*, 43 (2001), 1203–1216.

42. G. Bellanger and J. J. Rameaua, Corrosion of titanium nitride deposits on AISI 630 stainless steel used in radioactive water with and without chloride at pH 11, *Electrochimica Acta*, 40 (1995), 2519–2532.

43. S. R. Allah-karam, V. Vasantasree, and M. G. Hocking, Electrochemical potential mapping of a rapidly solidified processed light alloy, *Corrosion Science*, 43 (2001), 1645–1656.

44. H. N. McMurry, S. R. Magill, and B. D. Jeffs, Scanning reference electrode technique as tool for investigating localised corrosion phenomena in galvanized steels, *Ironmaking and Steelmaking*, 23 (1996), 183–188.

45. H. N. McMurry, Localized corrosion behaviour in aluminium–zinc alloy coatings investigated using the scanning reference electrode technique, *Corrosion*, 57 (2001), 313–322.

46. H. S. Isaacs, The use of the scanning vibrating electrode technique for detecting defects in ion vapour-deposited aluminium on steel, *Corrosion*, 43 (1987), 594–598.

47. G. Grundmeier, W. Schmidt, and M. Stratmann, Corrosion protection by organic coatings: electrochemical mechanism and novel methods of investigation, *Electrochimica Acta*, 45 (2000), 2515–2533.

48. M. F. Montemor, A. M. Simoes, and M. J. Carmezim, Characterization of rare-earth conversion films formed on the AZ31 magnesium alloy and its relation with corrosion protection, *Applied Surface Science*, 253 (2007), 6922–6931.

49. R. S. Lillard, P. J. Moran, and H. S. Isaacs, A novel method for generating quantitative local electrochemical impedance spectroscopy, *Journal of the Electrochemical Society*, 139 (1992), 1007–1012.

50. J. N. Murray, Electrochemical test methods for evaluating organic coatings on metals: an update: Part III. Multiple test parameter measurements, *Progress in Organic Coatings*, 31 (1997), 375–391.

51. V. Barranco, N. Carmona, J. C. Galvan, M. Grobelny, L. Kwiatkowski, and M. A. Villegas, Electrochemical study of tailored sol–gel thin films as pre-treatment prior to organic coating for AZ91 magnesium alloy, *Progress in Organic Coatings*, 68 (2010), 347.

52. D. O. Wipf and A. J. Bard, Scanning electrochemical microscopy: 10. High-resolution imaging of active-sites on an electrode surface, *Journal of the Electrochemical Society*, 138 (1991), L4–L6.

53. D. O. Wipf and A. J. Bard, Scanning electrochemical microscopy: 7. Effect of heterogeneous electron-transfer rate at the substrate on the tip feedback current, *Journal of the Electrochemical Society*, 138 (1991), 469–474.

54. Y. Y. Zhu and D. E. Williams, Scanning electrochemical microscopic observation of a precursor state to pitting corrosion of stainless steel, *Journal of the Electrochemical Society*, 144 (1997), L43–45.

55. M. Shao, Y. Fu, R. Hu, and C. Lin, A study on pitting corrosion of aluminum alloy 2024-T3 by scanning microreference electrode technique, *Materials Science and Engineering A*, 344 (2003), 323–327.

56. Y. Zuo, H. Wang, J. Zhao, and J. Xiong, The effects of some anions on metastable pitting of 316L stainless steel, *Corrosion Science*, 44 (2002), 13–24.

57. K. Hladky and J. L. Dawson, The measurement of pitting corrosion using electrochemical noise, *Corrosion Science*, 21 (1981), 317–322.

58. K. Hladky and J. L. Dawson, The measurement of corrosion using electrochemical $1/f$ noise, *Corrosion Science*, 22 (1982), 231–237.

59. D. A. Eden, K. Hladky, D. G. John, and J. L. Dawson, Electrochemical noise resistance, Paper 274, presented at Corrosion '86, NACE, Houston, TX, 1986.

60. F. Mansfeld and H. Xiao, Electrochemical noise analysis of iron exposed to NaCl solutions of different corrosivity, *Journal of the Electrochemical Society*, 140 (1993), 2205.

61. U. Bertocci, C. Gobrielli, F. Huet, and M. Keddam, Noise resistance applied to corrosion measurements: I. Theoretical analysis, *Journal of the Electrochemical Society*, 144 (1997), 31–37.

62. R. A. Cottis, S. Turgoose, and R. Newman, *Corrosion Testing Made Easy: Electrochemical Impedance and Noise*, NACE International, Houston, TX, 1999.

63. R. A. Cottis, Interpretation of electrochemical noise data, *Corrosion*, 57 (2001), 265–285.

64. Y. J. Tan, S. Bailey, and B. Kinsella, The monitoring of the formation and destruction of corrosion inhibitor films using electrochemical noise analysis, *Corrosion Science*, 38 (1996), 1681.

65. W. P. Iverson, Transient voltage changes produced in corroding metals and alloys, *Journal of the Electrochemical Society*, Electrochemical Science (1968), 617–618.

66. Y. J. Tan, Sensing localized corrosion by means of electrochemical noise detection and analysis, *Sensors and Actuators B*, 139 (2009), 688–698.

67. P. C. Searson and J. L. Dawson, Analysis of electrochemical noise generated by corroding electrodes under open-circuit conditions, *Journal of the Electrochemical Society*, 135 (1988), 1908.

68. Y. J. Tan, B. Kinsella, and S. Bailey, An experimental comparison of corrosion rate measurement techniques: weight-loss measurement, linear polarization, electrochemical

impedance spectroscopy and electrochemical noise analysis, *Proceedings of Corrosion and Prevention '95*, Australasian Corrosion Association, Perth, WA, Australia, 1995.

69. F. Mansfeld and Z. Sun, Technical note: Localization index obtained from electrochemical noise analysis, *Corrosion*, 55 (1999), 915–918.

70. M. Hashimoto, S. Miyajima, and T. Murata, An experimental study of potential fluctuation during passive film breakdown and repair on iron, *Corrosion Science*, 33 (1992), 905–915.

3

Visualizing Localized Corrosion Using Electrochemically Integrated Multielectrode Arrays

Electrode inhomogeneity and electrochemical heterogeneity are ubiquitous natural phenomena that are the root causes of nonuniform electrode processes, including localized corrosion. Currently, the prediction and prevention of localized corrosion such as pitting remain the most difficult, yet fundamentally important issues in corrosion science and engineering [1–4]. Although localized corrosion, including pitting, has long been recognized to be an extremely insidious and dangerous corrosion problem, effective localized corrosion control remains nonexistent. Despite the significant amount of data reported in the literature, the pitting process, especially its initiation stage, remains the subject of much debate, with many questions left unanswered [1–4]. For example, the most widely used mechanism of localized corrosion initiation on passive metals (i.e., passive film breakdown) remains hypothetical since pitting initiation from passive film breakdown has not been observed experimentally. Indeed, although research carried out over the past several decades has identified many interesting characteristics of pitting corrosion, experimental observation of critical processes such as pit nucleation remains difficult.

Heterogeneous Electrode Processes and Localized Corrosion, First Edition. Yongjun Tan.
© 2013 John Wiley & Sons, Inc. Published 2013 by John Wiley & Sons, Inc.

The most common methods of detecting localized corrosion are visual, optical, and scanning electron microscopic observation of corroded metal coupons. Corrosion coupon testing is obviously the simplest and most widely used method of studying corrosion processes. A metal coupon of known metallurgy, size, shape, and weight is exposed to a corrosive environment and is inspected for corrosion after a period of time (e.g., every 90 days). Visual inspection of a metal surface can often determine the sizes and shapes of a corroded area, such as pits, while a more detailed examination of a corroded surface can be carried out using optical and scanning microscopy. Corrosion products may be analyzed by methods such as SEM, energy-dispersive x-ray spectroscopy (EDS), x-ray diffraction (XRD), and infrared or Raman spectroscopy. Corrosion coupons are excellent sources of corrosion information if monitoring is continuous and results are well accumulated and documented. They are widely used in laboratory and field corrosion testing, monitoring, and inspection to determine "cumulative" corrosion damage, material thinning, and localized forms of corrosion such as pitting, crevice corrosion, weld- and heat-affected zone corrosion, and erosion corrosion. However, corrosion coupon tests have well-known limitations: They require long exposure periods to generate test results and require periodic removal of test specimens from the corrosive environment, which is cumbersome and may alter the progress of localized corrosion. They detect only the cumulative corrosion damage at the end of the exposure period and provide little information on specific events that may have triggered this damage. In general, corrosion coupon methods are unable to monitor or visualize localized corrosion processes on an instantaneous basis.

An electrical resistance probe is often referred to as an "intelligent" weight-loss coupon test. The electrical resistance probe detects corrosion by measuring the electrical resistance of a thin metal test wire (sensor element) since the resistance of the wire increases as the wire becomes thinner due to corrosion dissolution. An advantage of the electrical resistance probe is that it provides cumulative metal-loss values without the need to remove samples from the service environment. Another advantage of electrical resistance probes is that the technique is applicable to both conductive and nonconductive corrosion environments. A major disadvantage of the electrical resistance technique is that it is unable to detect localized corrosion since localized corrosion may not lead to significant metal dissolution or to noticeable change in electric resistance. The field signature method is a corrosion testing technique developed for monitoring corrosion in pipe-shaped structures. In this method a current is impressed through several electrodes (sensing pins) on the outer surface of a pipe, and the potential drops between these electrodes are measured. The sensing pins are typically separated by a distance of two to three times the wall thickness. Changes in the geometry in the form of general corrosion, erosion corrosion, cracking, or pitting will disturb the potential field in the metal. The change in the potential field pattern measured is related to a change in pipe wall thickness, and thus the development of corrosion or cracks can be detected. This technique is nonintrusive and is used to measure corrosion damage over a relatively large section of a structure. Possible limitations of the technique include its limited resolution and relatively slow response to corrosion. It was reported to take one year to respond to a corrosion rate of 1 m/yr [5].

Ultrasonic testing utilizes high-frequency sonic waves that transverse the thickness of a specimen and return to the probe to determine the thickness of the specimen. Ultrasonic tests are used widely for field inspection of cracks and localized corrosion damage. Automatic ultrasonic scanning and recording techniques combined with computer techniques are able to produce three-dimensional maps of the corroded surface. However, this technique responds relatively slowly to corrosion damage, and it was reported to take four years to respond to a corrosion rate of 1 m/yr [5]. Radiography makes use of the penetrating quality of shortwave electromagnetic beams, which may be x-rays or γ-rays, to image corrosion and to determine pit depths and the degree of thinning due to corrosion. Radiography is limited to detecting narrow cracks that are oriented in directions other than parallel to the electromagnetic beams. Techniques such as eddy current are used to visualize initial cracks caused by stress corrosion or corrosion fatigue. The eddy current technique detects surface cracks, pits, or other defects by measuring disturbed eddy currents on a material surface. However, the eddy current is restricted to a small layer in the surface of the metal called the *skin depth*. In industrial applications, equipment known as a "pig" (pipeline inspection gauge) has been developed based on these techniques and their combination for internal examination and the visualization of corroded pipes and tubes. The pig follows the flowing medium in the pipeline and records corrosion-related data for analysis after it is removed from the pipe. These corrosion monitoring or visualization techniques are obviously very useful in corrosion inspection; however, they are able to detect corrosion only when sufficient corrosion has occurred to cause an accumulated change in the bulk material properties.

In the real world, corrosion rates always change with time, and thus it is important to visualize corrosion, especially localized corrosion processes, on a continuous basis. Electrochemical methods such as corrosion potential measurement, potentiodynamic polarization, linear polarization resistance (LPR), electrochemical impedance spectroscopy (EIS), and electrochemical noise analysis (ENA) can be fast, sensitive, and versatile corrosion testing techniques, but they have limitations in visualizing localized corrosion and in measuring localized corrosion kinetics [7,8]. They are usually applied only to the measurement of uniform corrosion rates, the determination of localized corrosion tendency, and the study of corrosion-related phenomena such as metal passivation. Scanning probes such as the scanning reference electrode technique (SRET) and the scanning vibrating electrode technique (SVET) have been used to detect and visualize localized corrosion by detecting ionic current flows in the electrolyte phase over an electrode surface; however, they measure only the currents in the solution phase and not exactly at the electrode–electrolyte interface, and thus the scanning results may not be able to reflect the complex distribution of ionic current sufficiently well on a nonuniform electrode surface or in a corroding pit. Scanning probes commonly operate in a relatively specific area; thus, the scan images may not necessarily represent the full details of electrode processes that involve different reactions occurring simultaneously over distinctively separated electrode areas. Scanning probes may also have difficulties in sensing corrosion on uneven surfaces under deposits or in crevices.

To characterize a heterogeneous electrochemical process such as localized corrosion, electrochemical parameters at local areas of a working electrode surface, such as local corrosion potential and electrochemical reaction current, have to be determined. Conventional electrochemical techniques commonly use a one-piece working electrode, an electrode that is constructed by a single piece of metal. When such an electrode is used, only mixed and averaged electrochemical parameters (e.g., a mixed potential over the entire electrode surface) is measurable. These measured electrochemical parameters are not related to either the anodic or the cathodic zone of a heterogeneous electrode surface. With a one-piece electrode, it is impossible to measure the galvanic currents that flow in the electrode body between localized anodic and cathodic sites, since an ammeter cannot be inserted between anodic and cathodic sites which are located on a single piece of metal surface. For these reasons, the conventional one-piece electrode has a major limitation in measuring and studying heterogeneous electrochemical processes.

One approach to overcoming this limitation is the development of an electrochemically integrated multielectrode array: the *wire beam electrode* (WBE) [6–12]. The WBE is a nonscanning probe technique that is able to visualize the processes of localized corrosion by measuring electrochemical parameters from local areas of a working electrode surface, such as local corrosion potential and galvanic current, providing spatial and temporal information on localized corrosion [6–24]. In this chapter we provide an overview of the WBE method and its applications in visualizing and characterizing electrochemical heterogeneity and localized corrosion. Several typical localized corrosion phenomena, including localized corrosion in an Evans water drop, are described to illustrate applications of the WBE method in visualizing localized corrosion processes, mechanisms, and prevention techniques [25–27].

3.1 AN ELECTROCHEMICALLY INTEGRATED MULTIELECTRODE ARRAY: THE WIRE BEAM ELECTRODE

The WBE is a unique multielectrode array that typically consists of electrochemically integrated and individually addressable electrodes. A metal wire array is constructed from many electrically insulated sensor elements that are made from metal wires of identical, dissimilar, galvanized, heat-treated, or stressed metal materials with the terminals of the wire bundle connected together. Figure 3.1 shows a conceptual design and three different configurations of a WBE. The surface shape of a WBE simulates the surface shape of a practical metal surface. The configuration of a WBE can be changed to meet various electrochemical and corrosion testing requirements; for example, it can be designed as a tube shape to study localized corrosion in a pipeline and as a rod shape to study the corrosion of embedded materials (see Figure 3.1). When a metal bundle is constructed from dissimilar metal

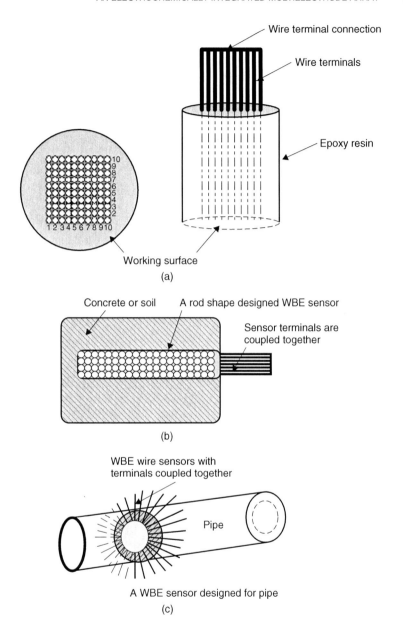

Figure 3.1 Conceptual design and various configurations of a WBE.

materials, the distribution of metal wires at the working surface should simulate the metal composition and distribution of a one-piece electrode surface.

The working surface of a WBE is integrated electrochemically by coupling all the terminals of the wires in the solid–phase and by closely packing all the wires in the solid/electrolyte interface. This electrochemical integration minimizes the

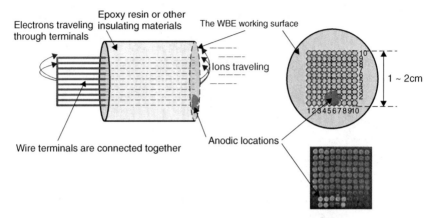

Figure 3.2 Working principle of a WBE system. (From [27].)

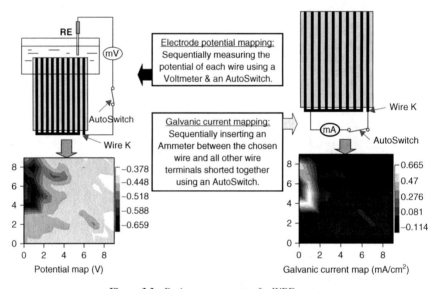

Figure 3.3 Basic measurements of a WBE system.

influence of the insulating layer on electron and ion movements, and thus the working surface of a WBE could effectively simulate a conventional one-piece electrode surface in electrochemical activity and behavior. Figure 3.2 illustrates the working principle of a WBE system, where localized corrosion is permitted to evolve dynamically and propagating freely and localized corrosion. Figure 3.3 illustrates the measurement of heterogeneous galvanic current and potential distributions from a WBE surface.

Similar corrosion patterns were observed experimentally over a WBE and large-area one-piece electrode surfaces when they were exposed to identical localized corrosion environments [7–12]. This can be explained theoretically:

1. When uniform corrosion occurs, there should be no difference in the macro- and microscales between localized areas of a large electrode surface. That is, a large-area electrode should behave in the same way as the parallel connection of many small individual minielectrodes (i.e., a WBE where all the wire terminals are connected together). The only difference between a large-area electrode and a WBE with the same total metal area is that the actual surface area of the WBE is larger because of the presence of an epoxy insulating layer between wires. This difference may affect the uniform corrosion process to some extent if the corrosion is controlled by diffusion, since the larger surface area of the WBE may have some influence on the radial diffusion process. However, in most cases, this influence should be negligible because the insulating epoxy layer is usually thin (approximately 20 μm in thickness). Therefore, it is expected that uniform corrosion occurs in a very similar way on the working surface of the WBE and on a large-area electrode–plate.

2. When localized corrosion occurs on the working surface of a WBE where all the wire terminals are connected together, electrons can move freely from anode to cathode through connected wire terminals, similar to the way they do in a large-area electrode. In this case, a possible difference between a WBE and a large-area electrode comes from the influence of the insulating layer between the wires in the WBE on the movement of ions between the localized anodes and cathodes. The insulating layer increases the distance of ion movement in electrolyte solution; however, if the solution resistance is not too large and the insulating layer is not too thick, this influence should be negligible or be deduced during data analysis. Another possible problem with a WBE is that the size of the wire may limit the propagation of localized corrosion in some cases. This problem could be overcome through a suitable choice of wire size. Thus, when localized corrosion occurs, it is also expected that the working surface of a WBE will effectively simulate a large-area electrode–plate. For these reasons, electrochemical integration of a WBE working surface permits electrochemical heterogeneity to evolve dynamically and propagate freely.

The choice of the surface area and cross section of each wire used in a WBE depends on corrosion and electrochemical testing conditions and objectives. For example, when a WBE is used to study pitting corrosion initiation, the surface area of each wire should be larger than the size of pits that are normally observed on large-area electrode or coupon surfaces at a certain stage of pitting propagation. The greater the number of wires, the better, provided that it is technically practical. The insulating layer between wires should normally be thin, and a large insulation resistance is essential. However, if the resistance of the electrolyte is low, a large space between neighboring wires is acceptable for the WBE surface to simulate a large-area electrode with reasonable accuracy [7].

A fundamental assumption on which the WBE concept is based is that the surface of each wire in an integrated WBE is homogeneous even if the overall WBE surface is heterogeneous. This is an approximation analogous to the principles of calculus: The surface area of each wire is much smaller than that of an integrated WBE surface; thus, it is reasonable to assume that each wire surface is homogeneous in electrochemical reaction rates and patterns [7,8]. This assumption allows electrochemical techniques and theories of describing uniform electrochemical processes to be applied to each wire in a WBE. That is traditional electrochemical theories and techniques are extended to study localized corrosion and other heterogeneous electrochemical processes. Based on this assumption and the Butler–Volmer equation, new equations describing the electrochemical kinetics of each wire have been derived and applied successfully to study the kinetics of localized corrosion processes [7,9].

Arrays of platinum, gold, or glassy carbon microelectrodes are commonly used in electroanalytical applications because they provide circumstances where a large and easily measured total current output is obtained while retaining many of the advantageous features possessed by the individual microelectrodes [28]. These widely spaced noble microelectrode arrays are usually designed to take advantage of three-dimensional diffusion for easy reaction current measurements; they are rarely used for the purpose of simulating or detecting electrode inhomogeneity or electrochemical heterogeneity over practical macroelectrode surfaces. The first application of an addressable multi-macroelectrode array to detect electrode inhomogeneities is probably in an experiment designed to determine inhomogeneity and its distribution in anticorrosion coatings [29].

The WBE was developed and used initially to detect local defects in organic coating films [29–31]. Its application has been extended to the study of crevice corrosion [6,10], the classical waterline corrosion process [7], and localized corrosion and inhibition [9–12]. The WBE and variously designed coupled multielectrode arrays have been utilized by many researchers in various interesting localized corrosion research applications [13–24]. Various names, such as multielectrode array, coupled electrode array, galvanically coupled multielectrode array, multichannel electrode, segmented electrode, and wire electrode, have been utilized to describe this type of electrode design. These studies have already shown that use of a WBE is a very promising method for visualizing and measuring localized electrode processes.

For example, Scully and co-workers [13–17] carried out experiments to investigate various forms of localized corrosion using coupled multielectrode arrays. They investigated and modeled spatial interaction among localized corrosion sites over coupled multielectrode array surfaces [13]. They studied the origins of persistent interactions among localized corrosion sites using AISI 316 stainless steel multielectrode arrays. Interactions between early dominating pits and the adjacent electrode surface were found to develop as regions of enhanced or suppressed pitting susceptibility. Oxide film alteration, mixed metal sulfide inclusion damage, and surface contamination were all considered to be possible origins of persistent interactions. Scully and co-workers carried out major research to understand the fundamentals of various corrosion phenomena using close-packed and widely spaced coupled multi electrode arrays [16]. It was found that widely spaced electrode arrays are optimized for high-throughput experiments capable of elucidating the effects of various variables on corrosion properties. For example, the effects of a statistical distribution of flaws on corrosion properties can be examined. Close-packed arrays enable unprecedented spatial and temporal information on the behavior of local anodes and cathodes. Interactions between corrosion sites can trigger or inhibit corrosion phenomena and affect corrosion damage evolution. They demonstrated that coupled multi electrode arrays allow investigations of the effects of small changes in solution chemistry on anode development. They were also able quantitatively to define conditions in which localized corrosion is initiated and spreads. They used close-packed coupled multielectrode arrays to simulate a planar electrode and to monitor the anodic current evolution as a function of position during initiation and propagation of crevice corrosion of AISI 316 stainless steel and Ni–Cr–Mo alloy 625 [17].

Fushimi et al. [18] studied galvanic corrosion of carbon steel welded with stainless steel with a multichannel electrode in which the welded specimen was divided into nine working electrodes. By detecting the spatial distribution of participating currents as a function of immersion time, they found that the weldment acted as a cathode, throughout the immersion period, while the other base steel became anodes or cathodes, depending on their location, immersion time, and the concentration of the electrolyte solution. They also investigated the ability of zinc-rich paint to protect the welded specimen as a sacrificial anode. Legat [19] studied the time and spatial evolution of steel corrosion in concrete with a coupled electrode array. It was found that microelectrode arrays can monitor the time and spatial evolution of steel corrosion in concrete. The currents measured reliably indicated the temporal anodic and cathodic activities of individual electrodes, and the assessment of general corrosion rates is also possible. Zheng et al. [20] used coupled electrode arrays to characterize chromate conversion coating formation and breakdown processes. Zhong and Zhang studied the anticorrosion performance of temporarily protective oil coatings [21,22]. Using WBE-based experimental techniques, they investigated electrochemical inhomogeneity in temporarily protective oil coatings by sensing the potential variation over a WBE surface coated with preventive oil films [21]. It was found that the distribution of corrosion potential on the surface of an oil-coated WBE was heterogeneous. When the oil film degrades, the distribution of corrosion potential was found to change from a normal probability distribution to a discontinuous binomial distribution.

A WBE was also used both to investigate the self-repairing ability of a temporarily protective oil coating and to investigate the anticontamination performance of temporarily protective oil coatings. It was shown that salt contamination on the metal substrate had an influence on the heterogeneous distributions of corrosion potential and polarization resistance [22]. Zhang et al. studied various forms of electrochemical inhomogeneity, such as galvanic corrosion in a zinc–steel couple [23]. Yang et al. attempted to apply the WBE in industrial applications [24].

3.2 VISUALIZING THE PROGRESSION OF LOCALIZED CORROSION IN AN EVANS WATER DROP

Water-drop corrosion [25] is a classical method of studying localized corrosion that illustrates the establishment of local anodes and cathodes due to the formation of oxygen concentration cells. Water-drop corrosion was first reported by Evans in 1926 by visual observation of color changes in a water drop [25]. It has been discussed in almost every textbook on corrosion as a classical experiment in corrosion science. In a water-drop corrosion experiment, a drop of NaCl solution with small additions of phenolphthalein and potassium ferricyanide is placed on a horizontal steel sheet. After a period of exposure, a large anode in the center of the drop surrounded by a ring of rust and a large cathode is observed based on the color changes in the solution. Chen and Mansfeld were the first to measure the potential distribution in a water drop using a Kelvin probe [32]. Water-drop corrosion also occurs in practical atmospheric corrosion conditions.

The WBE method has been used to visualize and elucidate the processes of water-drop corrosion and its cathodic protection by measuring electrochemical parameters such as corrosion potential and coupling current (galvanic corrosion) and their distributions directly from water-drop corrosion areas using a WBE. Figure 3.4 shows an experimental design for investigating water-drop corrosion using a WBE. The WBE used in this work [8] was made of 100 mild steel wires. These wires were embedded in an epoxy resin, insulated from each other by a very thin epoxy layer. Each wire had a diameter of 0.18 cm and acted as both

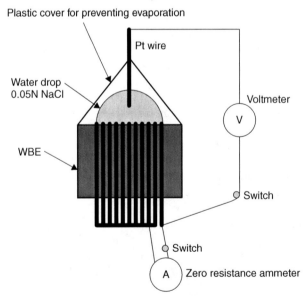

Figure 3.4 Water-drop corrosion experiment designed to measure galvanic corrosion current distribution and the system corrosion potential. (From [8].)

a minielectrode (sensor) and as a corrosion substrate. The working area (the area occupied by the wire beam) was approximately 3.2 cm^2 (1.8 cm $\times$ 1.8 cm). The distribution of these minielectrodes in the working surface of the WBE is shown in Figure 3.1. The working surface of the WBE was polished with 800- and 1200-grit silicon carbide paper and cleaned with ethanol and isopropanol. The working surface was exposed to the corrosive environment (a drop of 0.05 N NaCl solution) under static conditions at about $20°C$ to allow corrosion to occur. During exposure periods, all the wire terminals of a WBE were connected together to allow electrons to move freely between wires, similar to the manner used for mild steel electrodes and plates with a larger surface area. A platinum wire was used as a reference electrode for corrosion potential measurements since in this experiment a platinum wire is more convenient than other conventional reference electrodes. An automatic zero-resistance ammeter (AutoZRA, ACM Instruments, England) was used to record the corrosion potential of each wire and to measure the galvanic current flowing between each wire and the remaining wires of the entire wire bundle. The AutoZRA enables current (325 mA to 10 pA) and voltage to be measured accurately and to be recorded automatically. Its data logging software runs in a Microsoft Windows (Microsoft Corporation) environment, complete with a real-time Excel link. Thus, potential and current data analysis and plotting can be performed using Microsoft Excel.

The measurements of galvanic currents (I_{gk} for wire k of the wire bundle) flowing between each wire and the system were carried out using the experimental design shown in Figure 3.4. An AutoZRA performed the measurement on an individual wire terminal in sequence using an automatic switch. The system corrosion potential (E_{sys}) was also measured using this experimental design by employing a voltmeter against a platinum reference electrode. The corrosion potential of each wire (E_k for wire k of the wire bundle) was measured using the experimental design shown in Figure 3.5. When measuring the corrosion potential of wire k against the platinum reference electrode, the terminal of the wire selected was temporarily disconnected from the wire system and connected to the AutoZRA. There is an interval of 10 seconds between the measurement of successive wires. The measurements were repeated regularly during the experimental period. Visual observations were made at different stages of the exposure period.

Using the experimental designs shown in Figures 3.4 and 3.5, galvanic current and potential distributions in the water drop were measured. Measurements were repeated at various stages of the exposure period to study the processes of the water-drop corrosion. When a small water drop (approximately 7 mm in diameter) was used, as shown in Figure 3.6a, a potential distribution similar to that reported by Chen and Mansfeld [32] was recorded. The lowest potential values occurred in the center of the water drop and, correspondingly, positive galvanic currents occurred in the center of the drop (Figure 3.6b). These results indicate clearly that in the center of the droplet, the steel wires become anodic, and near the edge of the droplet, where the diffusion path of oxygen was shortest, the steel wires become cathodic. Visual observations of the WBE working surface corresponded well with those reported in the literature [25,32].

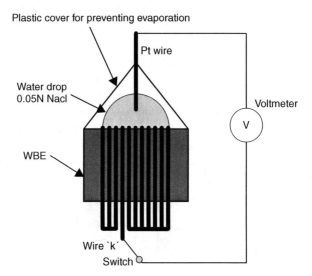

Figure 3.5   Water-drop corrosion experiment designed to measure the corrosion potential distribution. (From [8].)

However, when the size of the water drop was increased, the distributions of galvanic current and potential in the water drop changed significantly. Figure 3.7 shows the galvanic current distributions recorded from a larger water drop (approximately 12 mm in diameter) at various stages of exposure period. At the beginning of the exposure, as shown in Figure 3.7a, only a few isolated anodic sites, indicated by the positive galvanic currents, appeared among a large number of cathodic sites. In later stages (Figure 3.7b and c), some of the anodic sites merged together, forming a localized corrosion "ring" around the center of the water drop, unlike corrosion in the smaller water drop (Figure 3.6), where corrosion was concentrated in the center of the water drop. There was very good agreement between visual observations and galvanic current distribution (Figure 3.8). This result could be related to the change in the water drop's shape when its size was increased. The change in shape resulted in a more complex diffusion path of the oxygen and more complex distribution of the cathodic and anodic zones. Indeed, as the size of the water drop was increased further, a very different corrosion pattern was measured and observed (Figure 3.9). One side of the large drop became cathodic (negative current and less negative potential), while the remaining areas became anodic zones. It is interesting to note that the galvanic corrosion currents were distributed nonuniformly over the anodic zones. The areas close to the cathodic zone recorded higher anodic currents and thus more rapid corrosion. This is in agreement with visual observation (Figure 3.9c).

At the end of the water-drop corrosion experiment, a zinc wire was introduced into the water drop, replacing the position of wire 1 in the WBE. This zinc wire behaved as a sacrificial anode to prevent further water-drop corrosion. Protection current distribution was measured using the experimental design shown in

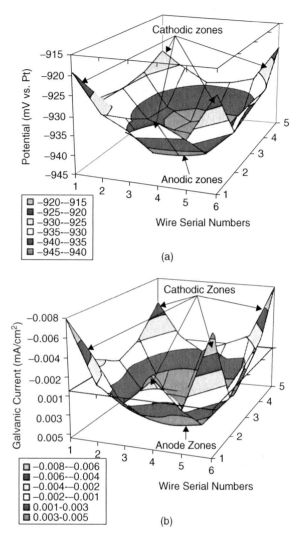

Figure 3.6   Corrosion potential (a) and galvanic current (b) distributions in a drop of 0.05 N NaCl solution (approximately 7 mm in diameter). (From [8].)

Figure 3.10. The zinc wire behaved as a sacrificial anode in this system and worked to prevent further water-drop corrosion. As shown in Figure 3.11, the zinc wire became the only anode in the WBE system and produced a large protection current (7.6 mA/cm^2) to prevent other mild steel wires in the water drop from further corrosion.

It is interesting to note that the protection currents were not distributed uniformly. As shown in Figure 3.11, larger protection currents appeared in a ring-shaped area that was previously a cathodic zone (Figure 3.7c). This is understandable since this ring-shaped area was close to the edges of the water

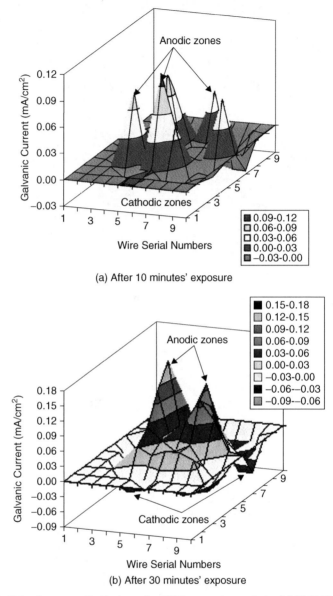

Figure 3.7 Galvanic current distributions of a WBE exposed to a drop of 0.05 N NaCl solution (approximately 12 mm in diameter). (From [8].) (*See insert for color representation of the figure.*)

drop, where the diffusion path of oxygen was shortest. An abundant supply of oxygen supported rapid cathodic reactions. This result suggests that the distribution of protection currents depends not only on the geometry of a metal structure and the conductivity of corrosion environments but also on factors such as the electrochemical heterogeneity of the metal surface. These factors need

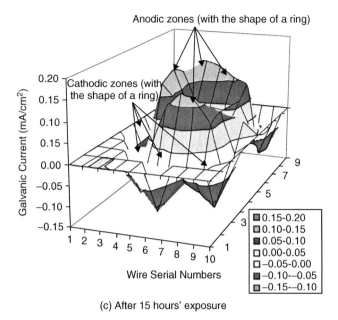

(c) After 15 hours' exposure

Figure 3.7 (*Continued*)

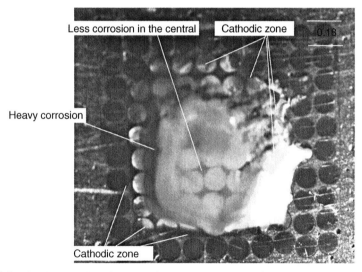

Figure 3.8 Visual observation of a WBE after exposure to a drop of 0.05 N NaCl solution (approximately 10 mm in diameter) for 17 hours. (From [8].) (*See insert for color representation of the figure.*)

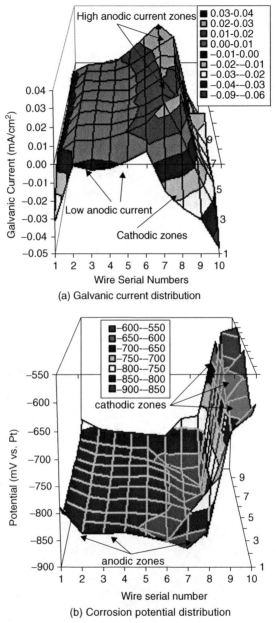

(a) Galvanic current distribution

(b) Corrosion potential distribution

Figure 3.9   Galvanic current (a) and potential (b) distributions and visual observation (c) of a WBE exposed to a drop of 0.05 N NaCl solution (approximately 19 mm in diameter) for 19 hours. (From [8].) (*See insert for color representation of the figure.*)

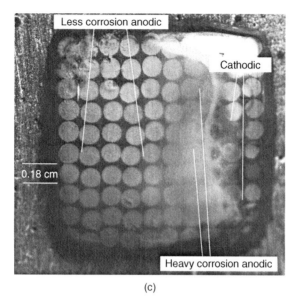

(c)

Figure 3.9 (*Continued*)

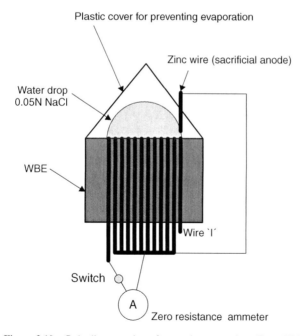

Figure 3.10 Cathodic protection of water-drop corrosion. (From [8].)

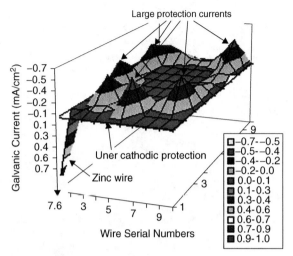

Figure 3.11 Cathodic protection current distributions over a WBE surface exposed to a drop of 0.05 N NaCl solution (approximately 12 mm in diameter). (From [8].)

to be considered when a cathodic protection system is modeled and calculated mathematically. However, this mathematic modeling and calculation can be very difficult, especially when a protected metal structure has a complex shape and the corrosion environment is complicated. The use of a WBE could be a much better method of characterizing a cathodic protection system.

3.3 VISUALIZING LOCALIZED CORROSION IN NONUNIFORM ENVIRONMENTS WITH ION CONCENTRATION GRADIENTS

Electrochemical corrosion processes are often affected by the rates at which reactants can be supplied to the electrode and products can be dispersed from it. Determination of the movements of electrochemically active species, such as ion diffusion, migration, and permeation, is important for understanding many localized corrosion processes. For example, corrosion of steel rebar in concrete is believed to be initiated by the penetration of corrosive species such as chloride ions and oxygen molecules, and controlled by the rates of mass transport of these corrosive species. Although mathematic modeling using differential equations, such as Fick's law of diffusion, is useful for predicting and explaining the diffusion behavior of ions in a practical concrete environment, mass transport in practical environments such as in a concrete structure is often very complex and could be affected by many factors, such as interactions between ions and time-dependent changes in ion diffusivity. An experimental method that can visualize the effects of electrochemically active species on corrosion processes should be useful in understanding localized corrosion.

An approach to developing a novel method for instantaneous determination of corrosive ion movement and its effects on electrochemical corrosion processes has been reported using a WBE as the monitoring tool [12]. Figure 3.12 illustrates an experimental design that was used to measure electrochemical parameters from a corroding WBE surface under a simulated diffusion–corrosion environment. The WBEs employed in this experiment included a mild steel WBE and a stainless steel (SS316L) WBE. The mild steel WBE was fabricated from 100 mild steel wires (0.18 cm in diameter) by embedding wires in epoxy resin. Its working area was approximately 3.24 cm^2 (1.8 cm × 1.8 cm), and the total metallic area was approximately 2.54 cm^2. The stainless steel WBE was made from 100 identical insulated stainless steel 316L wires. Each wire had a diameter of 0.15 cm and acted both as a minielectrode and as a corrosion substrate. The working area (the area occupied by the wire beam) was approximately 2.25 cm^2 (1.5 cm × 1.5 cm) and the total metallic area was approximately 1.77 cm^2. The working surfaces of the WBEs were polished with 400-, 800-, and 1000-grit silicon carbide paper and cleaned with deionized water and ethanol before being positioned horizontally facing up. The WBE surface was then covered by a piece of filter paper that was cut into the shape shown in Figure 3.12 and was wetted in distilled water. To create slow and gradual ion diffusion over the WBE surface, a pipette filled with solid chemicals such as FeCl$_3$, NaCl, or NiSO$_4$ was placed on the stick side of the filter paper (see Figure 3.12) close to wire 1 of the WBE. In this case, solid chemical from the pipette would gradually dissolve and diffuse away from the wire 1 location, creating concentration gradients over the WBE surface. During exposure, all the wire terminals of the WBE were connected together, so electrons would move freely between wires, simulating the electrochemical behavior of a one-piece electrode. All experiments were carried out under static conditions at air-conditioned room temperature (approximately 20°C).

Corrosion potential and current maps were measured repeatedly during exposure of the WBE surface to the ions' diffusion–corrosion environment. A corrosion potential distribution map was obtained by measuring the open-circuit potential of each wire (V_k for wire k) sequentially against a standard calomel reference electrode (SCE) using a voltammeter (GillAC, ACM Instruments, England) and a computer-controlled automatic switch device. The GillAC was designed to measure both voltage and current automatically. The same experimental setup was used to measure galvanic current distribution maps from the WBE surface. Galvanic current flowing between each wire (I_{gk} for wire k) and the entire short-circuited WBE system was measured by connecting the GillAC between each wire terminal chosen and all other wire terminals (shorted together). Alternate measurements of corrosion potential and galvanic current distribution maps were made during the experiment's duration (usually, 8 hours). Experimentally measured raw electrochemical data were transferred to analysis software written under a Mathcad environment (Mathcad Professional 7, MathSoft, Inc., Massachusetts, USA) for producing corrosion potential and galvanic current maps [12].

Figure 3.13 shows the corrosion potential and galvanic current distribution maps from a mild steel WBE exposed to an NiSO$_4$ diffusion–corrosion condition for

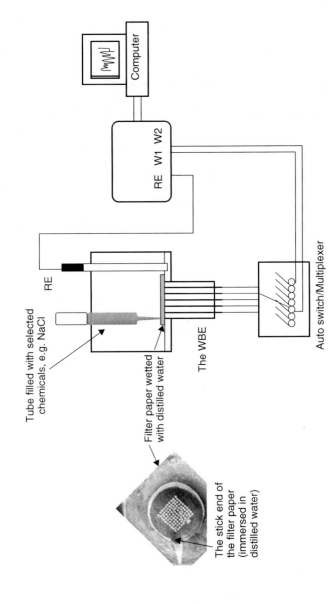

Figure 3.12 Experimental design for visualizing localized corrosion under a simulated diffusion–corrosion environment. (From [12].)

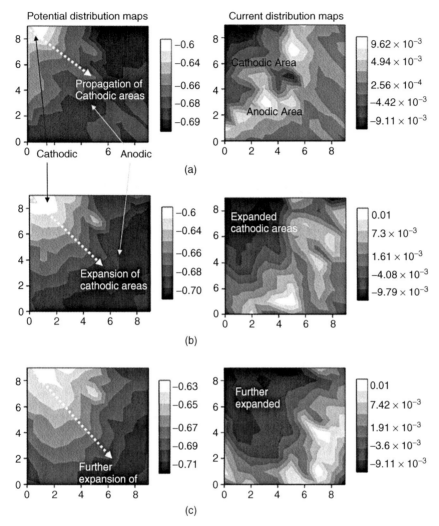

Figure 3.13 Corrosion potential and galvanic current distribution maps measured from a mild steel WBE exposed to $NiSO_4$ diffusion–corrosion conditions. (From [12].)

1, 3, and 5 hours. An interesting pattern observable in Figure 3.13 is that the cathodic area began at the wire 1 location and propagated along the diffusion direction. This behavior is understandable according to the Nernst equation since $NiSO_4$ diffusion would lead to a Ni^{2+} ion concentration gradient over the WBE surface and obviously would also lead to an electrochemical potential gradient. Indeed, as shown in Figure 3.13c, a potential distribution pattern appeared to emulate a diffusion pattern with a maximum potential difference of 80 mV across the WBE surface. The distinct separation of anodes and cathodes caused galvanic currents to flow from the cathodic sites to the anodic sites. The current distribution also appeared to emulate the diffusion pattern. These potential and current characteristics suggest that Ni^{2+} reduction ($Ni^{2+} + 2e^- \rightarrow Ni$, standard reduction potential: $-0.257V_{SHE}$) was the main cathodic reaction that controlled the corrosion process. The oxygen reduction reaction ($O_2 + 2H_2O + 4e^- \rightarrow 4OH^-$, standard reduction potential: $0.401V_{SHE}$) appeared to be a minor cathodic reaction due to the relatively low oxygen concentration.

Figure 3.14 shows corrosion potential and galvanic current distribution maps measured from a mild steel WBE exposed to $FeCl_3$ diffusion–corrosion conditions

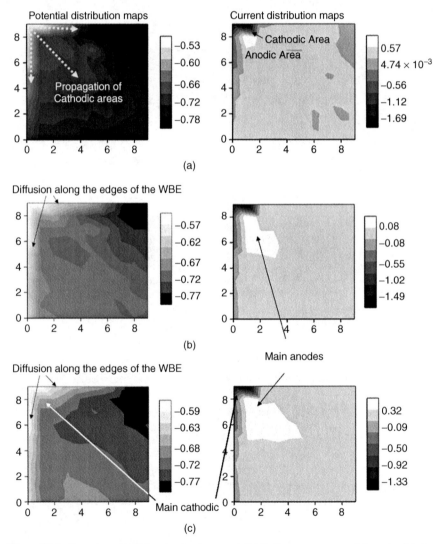

Figure 3.14 Corrosion potential and galvanic current distribution maps measured from a mild steel WBE exposed to $FeCl_3$ diffusion–corrosion conditions. (From [12].)

for 1, 3, and 5 hours. The general feature in Figure 3.14 is similar to that in Figure 3.13; that is, the potential and current distribution patterns appeared to emulate the ion diffusion pattern. However, the potential differences between the cathode and the anode in this system were much larger than those in the $NiSO_4$ system. As shown in the potential distributions of Figure 3.14, there was a maximum potential difference of 250 mV across the WBE surface. Such a large potential difference has driven significant galvanic currents to flow among wires in the WBE. As shown in the current distribution maps in Figure 3.14, the maximum anodic current (0.57 mA/cm^2) was 50 times higher than that in Figure 3.13. This galvanic current distribution pattern suggests that ferric ion reduction reaction ($Fe^{3+} + e^- \rightarrow Fe^{2+}$, standard reduction potential: $0.771 V_{SHE}$) was the dominant cathodic reaction. Its high standard reduction potential and high concentration at locations near wire 1 lead to significant and distinct separation of anodes and cathodes. However, it is not clear why the diffusion of $FeCl_3$ propagated mainly along the edges of the WBE and why the diffusion appeared to be slower than that of $NiSO_4$ diffusion (Figure 3.13).

Figure 3.15 shows the corrosion potential and galvanic current distribution maps measured from a mild steel WBE exposed to NaCl diffusion–corrosion conditions for 1, 3, and 5 hours. The features in Figure 3.15 are very different from those in Figures 3.13 and 3.14. At the early stage of exposure, as shown in Figure 3.15a, the distribution of cathodic and anodic sites appeared to be random and kept changing with time. With the extension of exposure, as shown in Figure 3.15b and c, higher cathodic currents was found to concentrate along the corners and edges of the WBE surface, with the magnitude decreasing in a contour-like manner toward the center of the WBE surface. This characteristic pattern appeared to be due to higher oxygen concentrations at these locations, where the oxygen supply was more sufficient through two- or three-dimensional diffusion. Higher oxygen concentrations along the edges and at the corners led to a higher potential at these locations (see Figure 3.15) and supported a higher rate of oxygen reduction reaction ($O_2 + 2H_2O + 4e^- \rightarrow 4OH^-$). In this case, oxygen concentration distribution was a dominant factor determining the corrosion pattern. The main effect of NaCl on corrosion was that NaCl diffusion led to a reduction in resistance of the filter paper and thus resulted in a continued increase in galvanic currents, which is clearly observable in the current distribution maps of Figure 3.15.

Significantly different features were observable from the corrosion potential and galvanic current distribution maps measured from SS316L WBEs exposed to $NiSO_4$ and NaCl diffusion–corrosion conditions. Typical maps from such systems are shown in Figures 3.16, where corrosion cathodes and anodes appeared to be distributed randomly; more important, these sites remained almost unchanged with the extension of diffusion–corrosion experiments. The characteristic concentration-controlled corrosion patterns shown in Figures 3.13 to 3.15 are totally absent in Figure 3.16. The most probable explanation of this phenomenon is that SS316L corrosion was controlled by the nonuniformity of the passive film on SS316L. It is well known that the corrosion resistance of stainless steel depends on the integrity and durability of its passive film. If passivity is not maintained and the

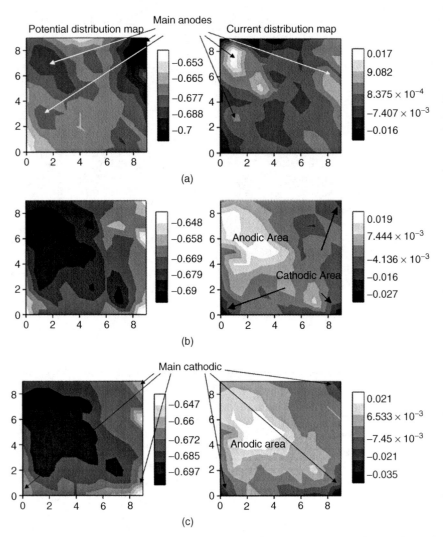

Figure 3.15 Corrosion potential and galvanic current distribution maps measured from a mild steel WBE exposed to NaCl diffusion–corrosion conditions. (From [12].)

passive film has localized weak sites or damage, these locations would become anodes where highly localized corrosion attack could occur. With the extension of the diffusion–corrosion experiment, as shown in Figures 3.16, potential at anodic locations shifted toward a more negative direction, which could be due to the continued deterioration of the passive film at anodic sites.

In general, the experimental results described above clearly demonstrate that the WBE method can be used to visualize localized corrosion processes and mechanisms that are affected significantly by mass transportation, such as ion diffusion.

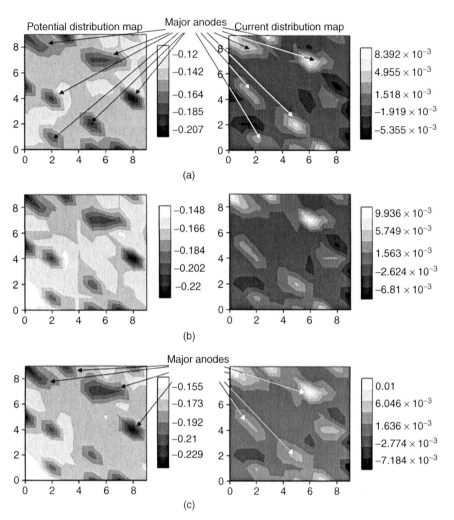

Figure 3.16 Corrosion potential and galvanic current distribution maps measured from an SS316L WBE exposed to $NiSO_4$ and NaCl diffusion–corrosion conditions. (From [12].)

3.4 VISUALIZING LOCALIZED CORROSION BY A WBE IN CONJUNCTION WITH SCANNING PROBES

The operation of electrochemical corrosion cells involves flows of ionic currents in the electrolytic phase and electronic currents in the metallic phase of a corroding metal surface. The determination of such ionic and electronic currents and their distribution over the metal–electrolyte interface is useful not only for understanding corrosion mechanisms but also for determining corrosion kinetics and predicting corrosion patterns. Such ionic and electronic currents are caused by potential variations over an electrochemically heterogeneous metal surface. Theoretically, their

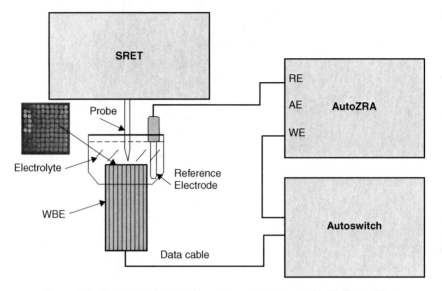

Figure 3.17 Experimental setup of WBE in combination with SRET. (From [33].)

distribution in the metallic and electrolytic phases is governed by the Laplace equation and Ohm's law and thus can be modeled mathematically. However, mathematical models developed to describe ionic and electronic current distributions are often difficult to verify experimentally, because, unfortunately, practical determination of such current distributions is experimentally difficult.

WBEs have also been used in conjunction with the scanning reference electrode technique (SRET) to visualize and study electrochemical corrosion processes and parameters from both the metallic and electrolytic phases of a corroding metal surface. The WBE method was used to map current and potential distribution in the metallic phase, and SRET was used to map current or potential distribution in the electrolytic phase. Figure 3.17 illustrates an experiment setup that combines the WBE and SRET measurement systems. The model SP100 SRET system (EG&G Company) consists of a scanning head, an electrolyte tank, a control electronics unit, and a WBE measurement system, which consists of a mild steel WBE, an automatic switch, and a measurement instrument (GillAC, ACM Instruments, England) [33]. The WBE was fabricated from 100 mild steel wires (0.18 cm in diameter) embedded in epoxy resin (see Figure 3.1). The total working area (the area occupied by the wire beam) was approximately 3.24 cm^2, and the metallic area was approximately 2.54 cm^2. The working surface of the WBE was polished with 400-, 800-, and 1000-grit silicon carbide paper and cleaned with deionized water and ethanol just before being positioned horizontally faceup in an electrolyte tank containing 250 mL of corrosive electrolyte (i.e., the Evans solution). The solution was prepared by dissolving 0.017 mol of NaCl and 0.008 mol of Na$_2$CO$_3$ in 1000 mL of deionized water as formulated by Evans [25]. Experiments were carried out under static conditions at about 20°C to allow corrosion to occur.

The WBE measurement system was used to map galvanic currents and corrosion potentials from the metallic phase. SRET measurements were carried out alternately with WBE measurements. To prevent the platinum probe tip from contacting the specimen surface during the SRET measurements, the WBE surface was adjusted with a spirit level to ensure that its surface was flat and vertical to the probe tip. The SP100 SRET system uses another platinum tip as the reference electrode. The distance between the probe tip and the WBE surface was adjusted carefully by moving the probe tip away from the specimen surface under computer control with the aid of a videocamera. The distance was adjusted to approximately 100 μm since the best resolution of the SP100 SRET system is achievable only when this distance is within the range 0 to 100 μm. The scanning ranges for the X and Y directions were set at 15 mm. Each SRET measurement produces a map containing 384 scan lines with 512 data points per line. WBE and SRET measurements were performed regularly, at certain desired intervals, during the duration of electrode exposure. During periods of non-measurement, all the wire terminals of the WBE were connected together to allow electrons to move freely between wires, in a manner similar to that used for single-piece mild steel electrodes with a larger surface area. Visual observations were made at various stages of the exposure period.

Localized corrosion in an Evans solution is a classical phenomenon that is often used in textbooks to illustrate the establishment of local anodes and cathodes [25]. Highly localized corrosion is often found to occur in the solution, leading to distinct separation of the anodic and cathodic areas on a steel surface. This experiment is traditionally carried out by visual observation of color changes in a water drop or at a waterline. This present work further elucidates localized corrosion processes in the solution, with and without the effects of solution pH changes and surface coatings, by measuring electrochemical parameters such as corrosion potential from both the metallic and solution phases.

Figure 3.18 shows typical results of WBE and SRET measurements from a mild steel WBE exposed to an Evans solution with a pH value of approximately 10.63. The WBE current and potential maps, which are typically as shown in Figure 3.18a and b, clearly correlate with each other. In particular, the anodic sites in the WBE maps (i.e., areas with positive current and more negative potential values) correlated well with the main anodic sites in the SRET map (Figure 3.18c), which shows negative voltage values. In this corrosion system, the main anodic sites remained almost unchanged during the entire period of experiment. It is evident that the corrosion pattern shown in Figure 3.18d correlates well with the WBE and SRET maps. This result clearly demonstrates localized corrosion due to macrocell electrochemical activities occurring on well-separated and stable locations. The WBE method detected electrons traveling continuously from the anode areas to the cathode areas through the electrode body, and at the same time the SRET detected ions traveling between anode and cathode areas through the electrolyte. However, by comparing Figure 3.18a, b, and c, it can be seen clearly that the cathodic sites detected by SRET differed from those detected using the WBE method. In Figure 3.18c, the SRET map shows a major cathodic zone located near the right side of the major

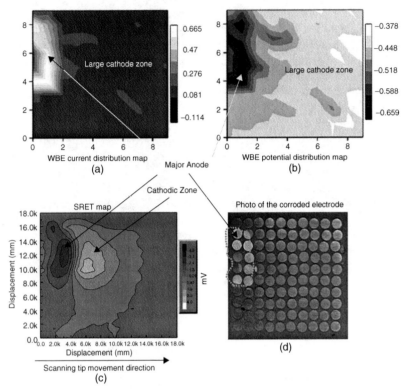

Figure 3.18 Typical results of WBE and SRET measurements from a mild steel WBE, exposed to an Evans solution. (From [33].)

anodic zone, while Figure 3.18a and b both indicate a large cathodic zone over the electrode surface. This is a phenomenon that has been observed repeatedly in other experiments, and it is believed to be due to the interference of SRET scanning on electrochemical corrosion processes. During SRET measurement, the scanning tip moved from left to right (as indicated in Figure 3.18c). This movement is expected to stir the solution and thus to enhance mass transportation over the electrode surface and to interfere with the steady-state anodic and cathodic reactions. As a result, at the time of scanning, the cathodic reaction near the right side of the anodic zone was accelerated significantly, and thus large cathodic currents were detected by the SERT instrument. This problem should be considered a limitation of the SRET method and probably other scanning probe techniques as well.

The typical features in Figure 3.18 were reproducible, although the exact corrosion pattern was not always reproduced in repeated experiments. As shown in Figure 3.19, a repeat experiment produced six major anodes over a WBE surface exposed to the same Evans solution under the same experimental conditions. Obviously, the anodic and cathodic locations were very different from those in Figure 3.18; however, the characteristics of these anodes were similar to those observed in the previous experiment. They were very stable, and their location did

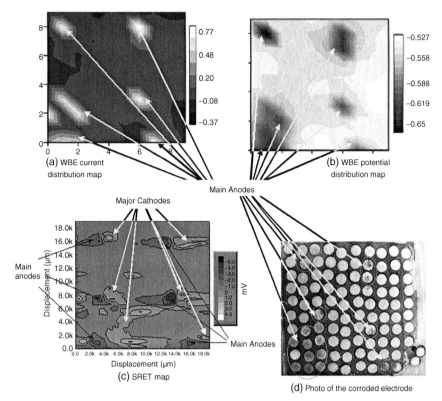

Figure 3.19   WBE galvanic current and potential distribution maps and SRET maps measured from a mild steel WBE surface after about 10 hours' immersion in Evans solution. (From [33].)

not change during the entire experimental period. The correlation between WBE, SRET, and visual observation in Figure 3.19 was good, although the WBE map appears to show more details of the local anodic and cathodic sites. The SRET map (Figure 3.19c) shows major cathodic zones located near the right side of the major anodic zones, which differ from the cathodic sites detected by the WBE method (Figures 3.19a and 3.19b, probably for the reason described above.

Corrosion behavior became very different when the pH of the Evans solution was adjusted from the original 10.63 to lower values by gradually adding HCl acid. Figure 3.20a to g show typical WBE and SRET maps measured at various stages of solution pH adjustments. When the pH of the solution was adjusted from 10.63 to 10.03, as shown in Figure 3.20a, the WBE and SRET measurements both detected one major anode. The anodic current density at the anodic site was high (the maximum was 2.06 mA/cm^2). With the extension of exposure, as shown in Figure 3.20b and c, the number of anodic sites increased considerably and the anodic current density decreased, suggesting that corrosion became more general and less concentrated. When the solution pH was decreased to 8.87, as shown in Figure 3.20d, more wires in the WBE became anodes and the maximum galvanic

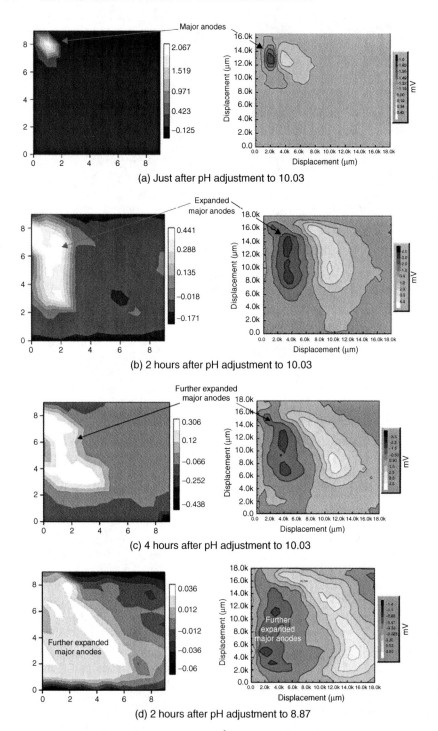

(a) Just after pH adjustment to 10.03

(b) 2 hours after pH adjustment to 10.03

(c) 4 hours after pH adjustment to 10.03

(d) 2 hours after pH adjustment to 8.87

Figure 3.20 WBE galvanic current (left; in mA/cm^2) and SRET and maps measured from a mild steel WBE surface exposed to Evans solution before and after pH adjustments. (From [33].)

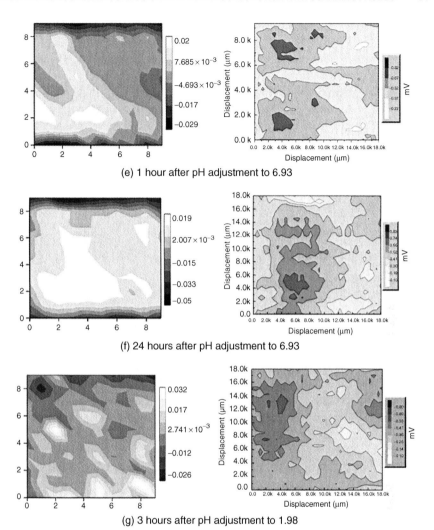

(e) 1 hour after pH adjustment to 6.93

(f) 24 hours after pH adjustment to 6.93

(g) 3 hours after pH adjustment to 1.98

Figure 3.20 (*Continued*)

current density kept decreasing. Obviously, the corrosion pattern changed from localized to general. The correlation between WBE, SRET, and visual observation was apparent, although the WBE map appears to show more details on local anodic and cathodic sites. Similar to observations made in previous experiments, the SRET maps show major cathodic zones located near the right side of the major anodic zones, which are different from the cathodic sites detected by WBE, probably also due to interference by the SRET scan on electrochemical corrosion processes.

With the pH decreased to 6.93, as shown in Figure 3.20e and f, the corrosion pattern became more and more general. Higher cathodic currents were recorded along the WBE edges, with the magnitude decreasing in a contour-like manner toward the center of the WBE surface. This pattern suggests oxygen concentration–controlled general corrosion behavior due to higher oxygen concentrations at corners and edges where oxygen supply was more abundant through two- or three-dimensional diffusion. In this case, oxygen concentration distribution was a dominant factor determining the corrosion pattern.

When the corrosion pattern became more general, the SRET and WBE maps became less well correlated with each other. This is understandable since separation of the anodic and cathodic areas on the steel surface became less distinctive when corrosion became more general. When the solution pH decreased further to 1.98, a major change occurred in the corrosion pattern: The distribution of anodic and cathodic sites became random (Figure 3.20g), which could indicate a corrosion mechanism change (i.e., the dominating cathodic reaction changed from oxygen reduction to hydrogen evolution). In this case, corrosion was no longer under oxygen diffusion control, due to an adequate supply of hydrogen ions. In such a system, the anodic and cathodic sites became small and not distinctly separated, and thus the correlation between the WBE and SRET maps was essentially lost. The WBE maps still appeared to visualize corrosion activities over the electrode surface. This result suggests that a combined WBE–SRET method is able to gain useful information on macrocell electrochemical corrosion processes that involve macroscale separation of anodes and cathodes. In such macrocell corrosion systems, maps measured using wbe and sret correlate with each other, and both methods were able to detect the locations of anodic sites.

Battocchi et al. studied the behavior of aluminum alloy 2024-T3 using a WBE in conjunction with scanning vibrating electrode technique (SVET) [34]. Two different types of nine-wire WBEs using (1) eight 1100 Al alloys and one pure copper wire and (2) nine wires of 2024 Al alloy have been built to emulate the behavior of a heterogeneous aluminum alloy surface where the current density at the surface is detected by SVET. The ability of SVET to obtain data with limited spatial resolution is used to determine if the nine-wire WBE configuration can simulate the behavior of a plate electrode. After immersion in a corrosive solution, the nine-wire electrode with a copper wire in the center showed the copper to be the cathode and the surrounding Al wires to be the anodes; this behavior clearly emulates the galvanic coupling of Cu and Al. In addition to pure Cu and pure Al coupling, they also used a WBE built using nine wires made of Al alloy 2024 based on the same manufacturing procedure. Very shortly after immersion in a corrosive solution, the WBE made of Al 2024 wires displayed anodic and cathodic currents distributed over the WBE surface, presumably related to the relative quantity of copper exposed on the wires' cross section to the solution [34]. With the extension of exposure, the number of wires showing anodic behavior decreased, suggesting that the wires were undergoing passivation. This trend continued until another wire becomes active and takes the place as the principal anode on the surface, perhaps reflecting a redistribution of the copper on the wire surfaces exposed to the electrolyte. The data acquired indicate that the behavior of a continuous surface under corrosion can be emulated using a multipiece WBE. Differences between the behavior of a plate and that of a wire electrode remain, and SVET showed a capability to address them [34].

REFERENCES

1. G. S. Frankel and N. Sridhar, Understanding pitting corrosion, *Materials Today*, 11 (2008), 38.

2. G. S. Frankel, Pitting corrosion of metals: a review of the critical factors, *Journal of the Electrochemical Society*, 145 (1998), 2186.

3. G. T. Burstein, et al., The origins of pitting corrosion, *Corrosion Engineering, Science, and Technology*, 39 (2004), 25.

4. M. P. Ryan, D. E. Williams, R. J. Chater, B. M. Hutton, and D. S. McPhail, Why stainless steel corrodes, *Nature*, 415 (2002), 770.

5. M. W. Joosten, K. P. Fischer, R. Strommen, and K. C. Lunden, Internal corrosion monitoring of subsea oil and gas production equipment, *Materials Performance*, April 1995, 44–48.

6. Y. J. Tan, A new method for crevice corrosion studies and its use in the investigation of oil-stain, *Corrosion*, 50 (1994), 266–269.

7. Y. J. Tan, Monitoring localized corrosion processes and estimating localized corrosion rates using a wire-beam electrode, *Corrosion*, 54 (1998), 403.

8. Y. J. Tan, Wire beam electrode: a new tool for studying localized corrosion and other heterogeneous electrochemical processes, *Corrosion Science*, 41 (1999), 229.

9. Y. J. Tan, S. Bailey, B. Kinsella, and A. Lowe, Mapping corrosion kinetics using the wire beam electrode in conjunction with electrochemical noise resistance measurements, *Journal of the Electrochemical Society*, 147 (2000), 530–540.

10. Y. J. Tan, S. Bailey, and B. Kinsella, Mapping non-uniform corrosion using the wire beam electrode method (I, II, and III), *Corrosion Science*, 43 (2001), 1905–1937.

11. Y. J. Tan, Corrosion science: a retrospective and current status, presented at the Electrochemical Society 201st Meeting, Philadelphia, G. S. Frankel, H. S. Isaacs, J. R. Scully, and J. D. Sinclair, Eds., 2002, PV2002-13.

12. N. N. Aung, Y. J. Tan, and T. Liu, Novel corrosion experiments using the wire beam electrode: II. Monitoring the effects of ions transportation on electrochemical corrosion processes, *Corrosion Science*, 48 (2006), 39–52.

13. T. T. Lunt, V. Brusamarello, J. R. Scully, et al., Interactions among pitting corrosion sites investigated with electrode arrays, *Electrochemical and Solid State Letters*, 3 (2000), 271–274.

14. T. T. Lunt, J. R. Scully, B. Brusamarello, A. S. Mikhailov, and J. L. Hudson, Spatial interactions among localized corrosion sites, *Journal of the Electrochemical Society*, 149 (2002), B163–B173.

15. N. D. Budiansky, J. L. Hudson, and J. R. Scully, Origins of persistent interaction among localized corrosion sites on stainless steel, *Journal of the Electrochemical Society*, 151 (2004), B233–B243.

16. N. D. Budiansky, F. Bocher, H. Cong, et al., Use of coupled multi-electrode arrays to advance the understanding of selected corrosion phenomena, *Corrosion*, 63 (2007), 537–554.

17. F. Bocher, F. Presuel-Moreno, and J. R. Scully, Investigation of crevice corrosion of AISI 316 stainless steel compared to Ni–Cr–Mo alloys using coupled multielectrode arrays, *Journal of the Electrochemical Society*, 155 (2008), C256–C268.

18. K. Fushimi, A. Naganuma, K. Azumi, and Y. Kawahara, Current distribution during galvanic corrosion of carbon steel welded with type-309 stainless steel in NaCl solution, *Corrosion Science*, 50 (2008), 903–911.

19. A. Legat, Monitoring of steel corrosion in concrete by electrode arrays and electrical resistance probes, *Electrochimica Acta*, 52 (2007), 7590–7598.

20. W. Zhang, B. Hurley, and R. G. Buchheit, Characterization of chromate conversion coating formation and breakdown using electrode arrays, *Journal of the Electrochemical Society*, 149 (2002), B357–B365.

21. Q. D. Zhong and Z. Zhang, Study of anti-contamination performance of temporarily protective oil coatings using wire beam electrode, *Corrosion Science*, 44 (2002), 2777–2787.

22. Q. D. Zhong, Potential variation of a temporarily protective oil coating before its degradation, *Corrosion Science*, 43 (2001), 317–324.

23. D. L. Zhang, W. Wang, and Y. Li, An electrode array study of electrochemical inhomogeneity of zinc in zinc/steel couple during galvanic corrosion, *Corrosion Science*, 52 (2010), 1277–1284.

24. L. T. Yang, N. Sridhar, C. S. Brossia, and D. S. Duna, Evaluation of the coupled multielectrode array sensor as a real-time corrosion monitor, *Corrosion Science*, 47 (2005), 1794–1809.

25. U. R. Evans, *An Introduction to Metallic Corrosion*, 3rd ed., Edward Arnold, London, 1981.

26. Y. J. Tan, Sensing pitting corrosion by means of electrochemical noise detection and analysis, *Sensors and Actuators B*, 139 (2009), 688–698.

27. Y. J. Tan, Sensing electrode inhomogeneity and electrochemical heterogeneity using an electrochemically integrated multi-electrode array, *Journal of the Electrochemical Society*, 156 (2009), C195–C208.

28. K. Aoki, Theory of microelectrodes, *Electroanalysis*, 5 (1993), 627–639.

29. Y. J. Tan, The effect of inhomogeneity in organic coatings on electrochemical measurements using a wire beam electrode: Part 1, *Progress in Organic Coatings*, 19 (1991), 89–94.

30. Y. J. Tan and S. T. Yu, The effect of inhomogeneity in organic coatings on electrochemical measurements using a wire beam electrode: Part 2, *Progress in Organic Coatings*, 19 (1991), 257–263.

31. C. L. Wu, X. J. Zhou, and Y. J. Tan, A study on the electrochemical inhomogeneity of organic coatings, *Progress in Organic Coatings*, 25 (1995), 379.

32. C. Chen and F. Mansfeld, Potential distribution in the Evans drop experiment, *Corrosion Science*, 39 (1997), 409–413.

33. Y. J. Tan, T. Liu, and N. N. Aung, Novel corrosion experiments using the wire beam electrode: III. Measuring electrochemical corrosion parameters from both the metallic and electrolytic phases, *Corrosion Science*, 48 (2006), 53–66.

34. D. Battocchi, J. He, G. P. Bierwagen, and D. E Tallman, Emulation and study of the corrosion behavior of Al alloy 2024-T3 using a wire beam electrode (WBE) in conjunction with scanning vibrating electrode technique (SVET), *Corrosion Science*, 47 (2005), 1165–1176.

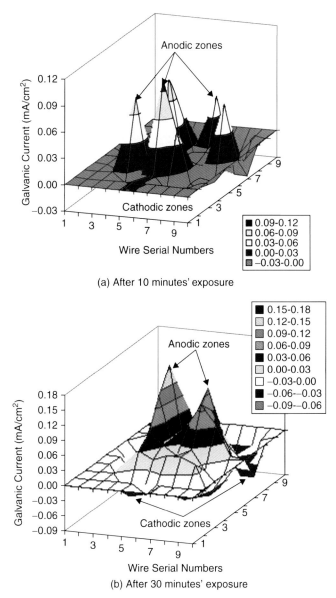

(a) After 10 minutes' exposure

(b) After 30 minutes' exposure

Figure 3.7 Galvanic current distributions of a WBE exposed to a drop of 0.05 N NaCl solution (approximately 12 mm in diameter). (From [8].)

Heterogeneous Electrode Processes and Localized Corrosion, First Edition. Yongjun Tan.
© 2013 John Wiley & Sons, Inc. Published 2013 by John Wiley & Sons, Inc.

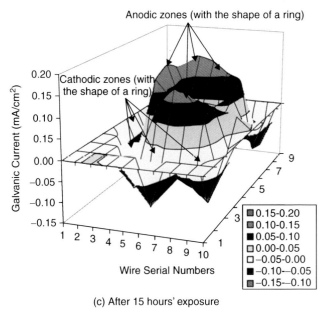

Anodic zones (with the shape of a ring)

Cathodic zones (with the shape of a ring)

Galvanic Current (mA/cm²)

0.20
0.15
0.10
0.15
0.00
−0.05
−0.10
−0.15

1 2 3 4 5 6 7 8 9 10

Wire Serial Numbers

9
7
5
3
1

■ 0.15-0.20
■ 0.10-0.15
■ 0.05-0.10
□ 0.00-0.05
□ −0.05-0.00
■ −0.10-−0.05
■ −0.15-−0.10

(c) After 15 hours' exposure

Figure 3.7 (*Continued*)

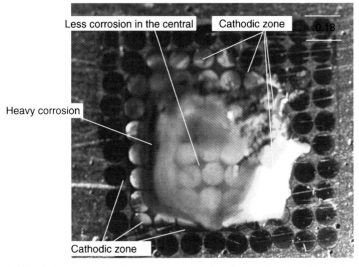

Less corrosion in the central   Cathodic zone   0.18

Heavy corrosion

Cathodic zone

Figure 3.8 Visual observation of a WBE after exposure to a drop of 0.05 N NaCl solution (approximately 10 mm in diameter) for 17 hours. (From [8].)

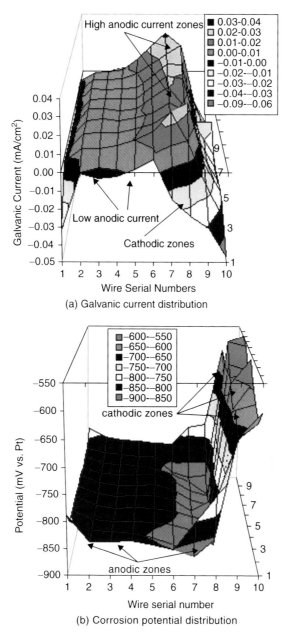

Figure 3.9 Galvanic current (a) and potential (b) distributions and visual observation (c) of a WBE exposed to a drop of 0.05 N NaCl solution (approximately 19 mm in diameter) for 19 hours. (From [8].)

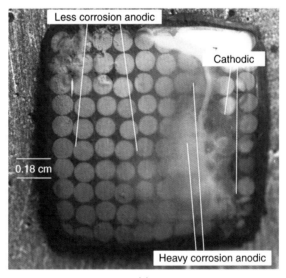

(c)

Figure 3.9 (*Continued*)

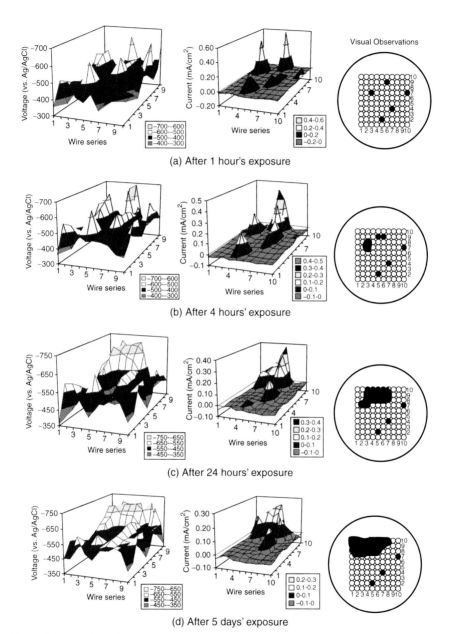

(a) After 1 hour's exposure

(b) After 4 hours' exposure

(c) After 24 hours' exposure

(d) After 5 days' exposure

Figure 4.6 Potential and galvanic current distributions over a WBE working surface during its exposure to 0.017 M NaCl + 0.008 M Na$_2$CO$_3$ solution. (From [9].)

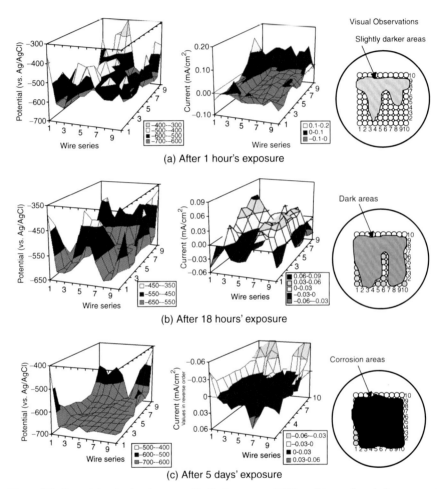

Figure 4.7 Potential and galvanic current distributions over a WBE working surface during exposure to 0.017 M NaCl solution. (From [9].)

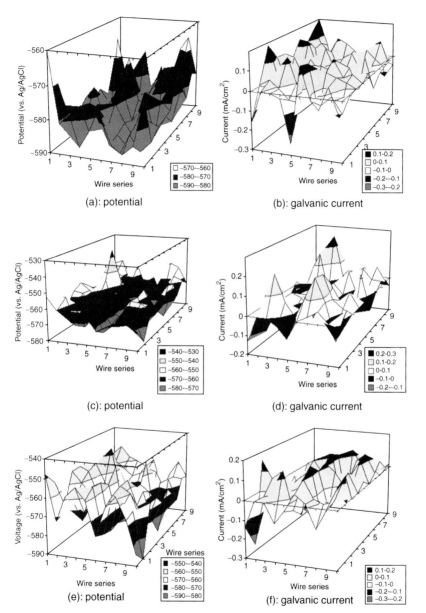

Figure 4.8 Potential and galvanic current distributions over a WBE surface after being exposed to 0.5 M H_2SO_4 solution. (From [9].)

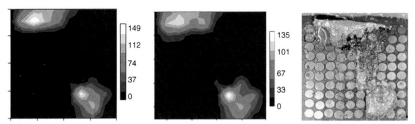

Calculated corrosion depth map   Observed corrosion depth map   Photo of corroded WBE

Figure 4.11   Corrosion depth maps calculated and observed microscopically and a photograph of the corroded surface after 406 hours' exposure (corrosion depths in μm) [12].

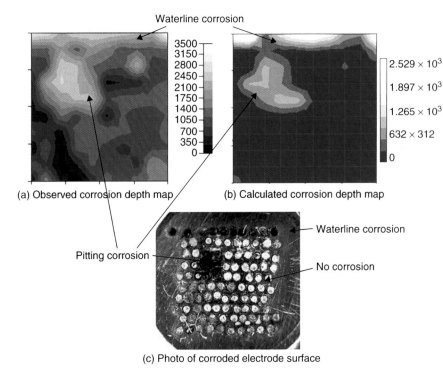

(c) Photo of corroded electrode surface

Figure 4.12   Corrosion depth maps and values (in μm) observed and calculated, plus a photograph showing a WBE probe surface after exposure to a waterline environment for 404 days.

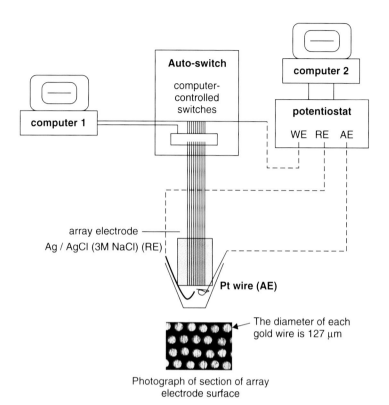

Figure 5.5 Experimental arrangement used to study the effects of electrode inhomogeneity on voltammetric responses. (From [7].)

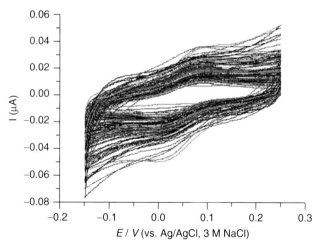

Figure 5.6 Cyclic voltammograms obtained from each 4,4′-dipyridyl disulfide–modified 127-μm-radius gold element (total of 98) of a gold multielectrode array, at a scan rate of 50 mV/s in 400μM cytochrome c (0.1 M NaCl in 20 mM phosphate buffer.) (From [7].)

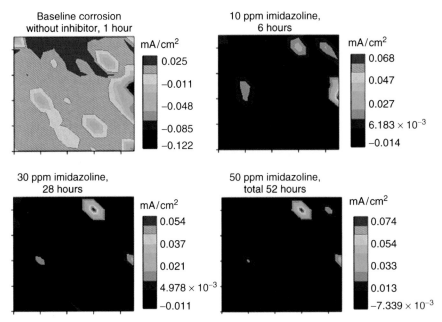

Figure 6.24 Galvanic current distribution maps measured over a WBE surface during 52 hours' exposure in brine with imidazoline present (current in mA/cm^2). (From [64].)

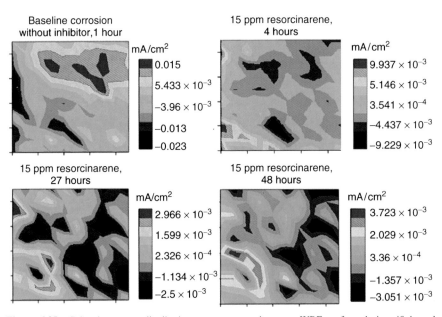

Figure 6.25 Galvanic current distribution maps measured over a WBE surface during 48 hours' exposure in brine with 15 ppm of resorcinarene acid present (current in mA/cm^2). (From [64].)

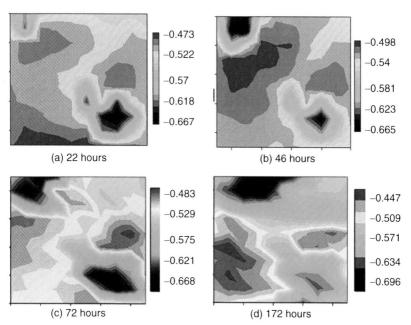

Figure 8.8 Corrosion potential distribution maps measured over a WBE surface during the first 172 hours' exposure (potential in volts)

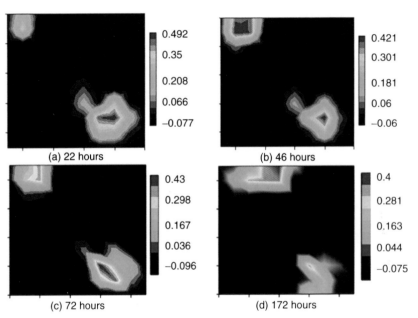

Figure 8.9 Corrosion current distribution maps measured over a WBE surface during the first 172 hours' exposure (in mA/cm^2).

4

Measuring Thermodynamic and Kinetic Parameters from Localized Corrosion Processes

The measurement of parameters relevant to the electrochemical thermodynamics and kinetics of a locally corroding electrode surface, such as the local electrode potential and local galvanic current, is required for determination of the susceptibility and rates of localized corrosion. Difficulties can be encounted when measuring localized electrochemical parameters by the conventional electrochemical method using a traditional macrodisk electrode. For example, the conventional method of measuring electrode potential using a one-piece electrode, a reference electrode, and a voltmeter only measures a potential that is a mixture of many local potentials, none of which we can evaluate independently using the Nernst equation or the Wagner–Traud mixed potential theory. When a one-piece electrode is used, it is impossible to measure the galvanic currents that flow between local anodic and cathodic sites in the metallic phase of an electrode since an ammeter cannot be inserted between the anodic and cathodic sites, which are located on a single piece of metal surface.

The only standard electrochemical method that is considered able to measure relative localized corrosion susceptibility is the pitting potential test [1]. This method involves anodic polarization of a specimen until localized corrosion is initiated, as indicated by a large increase in the current applied. An indication of

Heterogeneous Electrode Processes and Localized Corrosion, First Edition. Yongjun Tan.
© 2013 John Wiley & Sons, Inc. Published 2013 by John Wiley & Sons, Inc.

the susceptibility to initiation of localized corrosion is given by the potential at which the anodic current increases rapidly (i.e., the pitting potential). Conventional understanding is that the pitting potential is the potential above which pits are initiated. The nobler the pitting potential, obtained at a fixed scan rate in this test, the less susceptible is the alloy to the initiation of localized corrosion. However, it has also been found that at least in chloride solution, pits can nucleate and even grow at potentials below the pitting potential [2]. This suggests that the pitting potential does not necessarily mark the boundary between pitting and no pitting, although it is believed to separate the potential above which nucleated pits can propagate indefinitely to achieve a state of stable growth and below which stable growth cannot be achieved [2]. It should also be noted that the pitting potential test is not intended to correlate in a quantitative manner with the rate of localized corrosion or the distribution of pits.

Advanced physical techniques such as the scanning Kelvin probe (SKP) and scanning Kelvin probe force microscopy (SKPFM) have enabled mapping of the Volta potential by scanning the electric potential in the air over an electrode surface, just above the electrolyte. Schmutz and Frankel [3] identified a linear relationship between the Volta potential and the corrosion potential, indicating that the Volta potential is related to the thermodynamics of corrosion processes and thus could be used to determine the practical nobility of the surface. In a typical experimental study, de Wit used SKPFM to measure Volta potential differences over different parts of an aluminum alloy electrode, producing a potential map with high contrast that clearly identifies the locations of intermetallic particles [4]. It was found that the Volta potential differences between major intermetallics on the aluminum alloy—Al_7Cu_2Fe, $(Al,Cu)_6(Fe,Cu,Mn)$, and Mg_2Si—and the matrix were found to be about 300, 250, and -170 mV, respectively. However, it should be pointed out that the differences between the high and low Volta potential areas would not necessarily result in a large galvanic current flow [3,4] and that the Volta potential measured in the air would not equal the local corrosion potential measured in aqueous solution.

Advanced electrochemical scanning probe techniques such as the scanning reference electrode technique (SRET), scanning vibrating electrode technique (SVET), local electrochemical impedance spectroscopy (LEIS), and scanning electrochemical microscopy (SECM) have enabled the visualization of localized corrosion by mapping ionic currents from the electrolyte phase. However, as described in Chapter 2, these scanning probe techniques only detect the currents in solution phase, not exactly at the metal–solution interface; thus, they cannot accurately measure local corrosion currents flowing exactly at the metal–solution interface, and thus the current values measured are expected to be smaller than the actual corrosion current values.

As described in Chapter 3, the wire beam electrode (WBE) is able to measure and map electrochemical parameters from local areas of a working electrode surface, such as local corrosion potential and galvanic current from the metallic phases, providing spatial and temporal information on localized corrosion. However, further analysis of these corrosion potential and current data is required to determine the susceptibility and rates of localized corrosion. In this chapter we discuss the use of the WBE method, in conjunction with electrochemical noise analysis, in measuring localized corrosion thermodynamics and kinetics.

4.1 METHODS OF PROBING LOCALIZED CORROSION THERMODYNAMICS AND KINETICS

As described in Chapter 1, traditional electrochemical methods assume that the surface of a working electrode is electrochemically uniform and that the anodic and cathodic regions are distributed randomly. Such a simplification permits consideration of the entire working electrode surface simultaneously as an anode and a cathode, and thus offers certain advantages in calculations for uniform corrosion rates. However, this prerequisite condition does not apply to practical corrosion systems, where corrosion is often more or less localized. A practical electrode surface is normally characterized by the nonuniform distribution of local corrosion potential, local ohmic resistance, local solution composition and pH, and so on.

One approach to overcoming difficulties associated with traditional electrochemical techniques is the WBE concept. As shown in Figure 4.1, a WBE divides a

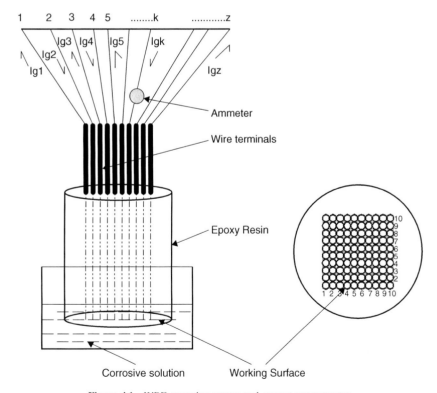

Figure 4.1   WBE corrosion system and current measurement.

nonuniform large electrode area into many small parts and allows the measurement of local potential and current. It is reasonable to assume that the surface of each wire is electrochemically uniform, in analogy to calculus in mathematics. This assumption allows electrochemical techniques and theories describing uniform corrosion to be applied to each wire in a WBE. This approach allows traditional electrochemical techniques to be extended to study localized corrosion processes. If all the terminals of a WBE are connected together, as shown in Figure 4.1, galvanic corrosion currents will flow between the wires. Some of the wires behave as anodes and others behave as cathodes. This is a multielectrode system where every wire is polarized by other wires and an overall system potential, E_{sys}, is achieved. The directions of galvanic currents, $I_{g1}, I_{g2} \cdots I_{gk} \cdots I_{gz}$, indicate whether a wire behaves in the system as an anode or a cathode and $\sum_{k=1}^{z} I_{gk} = 0$.

Localized electrode thermodynamics over the surface of any wire k selected, indicated by the local electrode potential, E_k, can be measured by disconnecting the wire temporarily from the WBE system and connecting it rapidly to a high-impedance voltmeter using a switching device. This E_k could, nonetheless, consist of components including an equilibrium potential E_k^0 over location k, where a miniequilibrium is supposed to exist, an ohmic drop, and an overpotential η_k that polarizes the wire k from E_k^0 to the overall potential of the integrated WBE system, E_{sys}. E_k^0 characterizes the thermodynamics of location k and can be determined by measuring E_k immediately after the rapid decay of the ohmic drop (usually in less than 10^{-12} second [5] and η_k, but before the occurrence of major changes in the surface minienvironment (due to relatively slow migration, diffusion, and convection processes). E_k^0 is a local equilibrium potential that is determined by local reactant and product concentrations of $[O]_k$ and $[R]_k$ and local temperature T_k over a steady-state WBE surface by the fundamental Nernst equation [6]. For a simple reversible electrode reaction, $O + ne^- \rightarrow R$, E_k^0 can be described as

$$E_k^0 = E^0 + \frac{RT_k}{nF} \ln \frac{[O]_k}{[R]_k} \qquad (4.1)$$

In practical cases such as electrochemical corrosion, however, electrode processes are usually more complicated, and two or more reactions can proceed simultaneously over an electrode surface (termed a *mixed electrode*). Traditionally, the Wagner–Traud mixed potential theory [7] is designed for such a mixed electrode under the assumption that all anodic and cathodic reactions are distributed uniformly and randomly over the entire electrode surface. Such a simplification permits consideration of the entire metal surface simultaneously as an anode and a cathode, and thus a uniform mixed potential distribution is achieved. This treatment offers certain advantages in traditional electrochemical thermodynamic and kinetic calculations; however, it does not apply to a heterogeneous electrode where there is a distinct separation of electrode reactions. The WBE concept allows the consideration of each wire surface as uniform and thus has a uniform mixed potential. The mixed potential over each location on the WBE surface can be measured independently and can be evaluated using the Wagner–Traud theory. In this way,

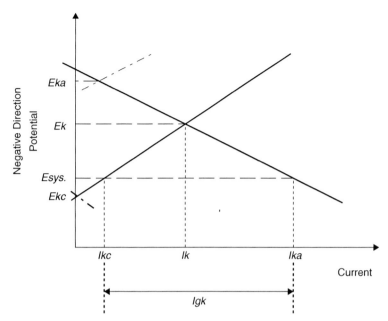

Figure 4.2   Evans plot indicating the polarization of wire k in a freely corroding WBE surface. (From [9].)

the nonuniform mixed potential distribution over the entire WBE surface can be determined and mapped.

An Evans diagram [8] is a simplified tool for analyzing the kinetics of mixed electrode processes such as localized corrosion under open-circuit conditions. Figure 4.2 shows an Evans plot that describes the polarization of wire k over a WBE surface under open-circuit free corrosion [9]. Here E_k and I_k are the open-circuit potential and corrosion current of wire k. E_k can be measured easily when wire k is disconnected from the multielectrode system using a voltmeter via a reference electrode. I_k can also be measured easily using traditional electrochemical techniques, such as linear polarization measurements, because corrosion on the wire surface can be assumed to be uniform. E_{kc} is the equilibrium potential of the cathodic corrosion reaction, and E_{ka} is the equilibrium potential of the anode corrosion reaction. E_{sys} is the overall corrosion potential of the entire WBE system, which can easily be measured using a voltmeter via a reference electrode. When wire k is connected to a multielectrode system, the anodic reaction current will be I_{ka}, and the cathodic reaction current will be I_{kc}. The difference between I_{ka} and I_{kc} is the galvanic (or coupling) current I_{gk}, which can be measured by inserting an ammeter into the multielectrode system (Figure 4.1). When wire k is connected to the multielectrode system, it is polarized by an overpotential $\eta_k = E_{sys} - E_k$. The overpotential is positive or negative, depending on whether the polarization is cathodic or anodic. The total cathodic current for wire k (I_{ka}) is a combination of the cathodic reaction current of an internal corrosion cell (equal to I_{kc}) and galvanic current (I_{gk}) (Figure 4.2). I_{kc} constitutes the uniform corrosion current component, and I_{gk} constitutes the localized corrosion current component of the wire corrosion. Uniform and localized corrosion thus occurs simultaneously at the minielectrode. The overall anodic corrosion reaction current of wire k, I_{ka}, can be written

$$I_{ka} = I_{gk} + I_{kc} \tag{4.2}$$

The anodic reaction current, I_{ka}, describes overall localized corrosion kinetics and could be correlated with traditional techniques, such as the optical measurement of pitting depth and the chemical analysis of ion concentration in local areas. If the anodic and cathodic reactions on the wires are activation controlled, corrosion on each wire can be considered to be uniform, and the open-circuit potentials of the wires and the potential of the entire system are far away from the equilibrium potentials of wires (E_{kc} and E_{ka} for wire k), the Butler–Volmer equation can be applied to wire k. At potential E_{sys}, the anodic reaction current can be written

$$I_{ka} = I_k \exp \left[\frac{2.3(E_{sys} - E_k)}{b_{ak}} \right] \tag{4.3}$$

Similarly, the cathodic reaction current can be written

$$I_{kc} = I_k \exp \left[\frac{-2.3(E_{sys} - E_k)}{b_{ck}} \right] \tag{4.4}$$

Substituting (4.3) and (4.4) in (4.2) gives

$$\frac{I_{gk}}{I_{ka}} = 1 - \exp \left[-\left(\frac{2.3}{b_{ak}} + \frac{2.3}{b_{ck}} \right) (E_{sys} - E_k) \right] \tag{4.5}$$

That is,

$$I_{ka} = \frac{I_{gk}}{1 - \exp \left[-\left(\dfrac{2.3}{b_{ak}} + \dfrac{2.3}{b_{ck}} \right) (E_{sys} - E_k) \right]} \tag{4.6}$$

The localized corrosion current, I_{ka}, can be estimated according to equation (4.3) or (4.6). Equation (4.6) is more convenient because when it is used there is no need to measure and calculate I_k. In equation (4.6), E_k can be determined simply by measuring the potential of wire k against a reference electrode while the wire k is disconnected (temporarily) from other wires. E_{sys} can be determined by measuring the WBE potential against a reference electrode while all the wire terminals are connected together. I_{gk} can be determined easily by inserting a zero-resistance ammeter in the E_{sys} measurement circuit. The Tafel slopes, b_{ak} and b_{ck}, can be estimated by performing a Tafel curve measurement. If the open-circuit potentials of wires in a WBE are very different from the potential of the entire system (i.e., the wires are under a high level of polarization), equation (4.6) can be simplified:

$$I_{ka} = I_{gk} \tag{4.7}$$

The criterion that permits the simplification of equation (4.6) to (4.7) could be that the difference between the open-circuit potential of a wire and the potential of the entire system is larger than 100 mV, an analog to the validity condition of the Tafel equation. Equation (4.7) suggests that when an electrode surface is under a high degree of polarization, the galvanic currents can be taken as a good estimation of localized corrosion currents. Localized corrosion currents on all locations of the WBE (wires 1 to z) can be estimated using equation (4.3), (4.6), or (4.7), and thus the corrosion rate distribution over the WBE surface can be mapped. Corrosion maps such as a three-dimensional spatial plot can then be produced to show localized corrosion rates and their distribution over the electrode surface [9].

However, if the WBE is under anodic or cathodic polarization, a more general analysis based on the Butler–Volmer equation [10,11] and the mixed potential theory [7] should be carried out [12]. If location k in a WBE is polarized from a mixed potential ($E_{\text{mix},k}$) to a preset polarization potential (E_{set}), the Butler–Volmer equation can be shown as

$$i_k = i^0_{\text{mix},k} \left[\exp\left[\frac{-\beta_k n F (E_{\text{set}} - E_{\text{mix},k})}{R T_k} \right] - \exp\left[\frac{(1 - \beta_k) n F (E_{\text{set}} - E_{\text{mix},k})}{R T_k} \right] \right] \tag{4.8}$$

where $i^0_{\text{mix},k}$, T_k, and β_k are the local electrochemical reaction current, temperature, and symmetry coefficient over a presumed uniform location k; n, F, and R are the charge involved, Faraday's constant, and the universal gas constant, respectively. This approach effectively extends conventional electrochemical kinetic methods developed based on the Butler–Volmer equation, such as the Tafel polarization and Stern–Geary linear polarization methods, to the determination of heterogeneous electrochemical kinetics. In equation (4.8), $E_{\text{set}}-E_{\text{mix},k}$ is the externally applied overpotential that polarizes the mixed electrode. For a free electrochemical corrosion process, the E_{set} equals the overall potential of the integrated WBE corrosion surface, E_{sys}, and $E_{\text{mix},k}$ equals the corrosion potential of location k, $E_{\text{corr},k}$. Equation (4.8) can be simplified into a special form for use in a free corrosion system:

$$I_{ka} = \frac{I_{gk}}{1 - \exp\left[-(2.3/b_{ak} + 2.3/b_{ck})(E_{\text{sys}} - E_{\text{corr},k}) \right]} \tag{4.9}$$

where I_{ka}, b_{ak}, and $b_{c,k}$ are the anodic corrosion dissolution current and the anodic and cathodic Tafel slopes over the wire k surface, respectively. I_{gk} is the galvanic current that flows into or out of wire k, which is measurable using an ammeter. Equations (4.9) and (4.6) are identical, suggesting that Evans plots and the

Butler–Volmer equation and mixed potential theory are both applicable in this analysis.

At any time t, the corrosion current at location k, $I_{corr,k}(t)$, can be calculated using equation (4.9), employing overpotential over a WBE surface, $E_{sys}(t) - E_{corr,k}(t)$, and galvanic current, $I_{gk}(t)$,

$$I_{corr,k}(t) = I_{ka}(t) = \frac{I_{gk}(t)}{1 - \exp\left[-(2.3/b_{ak} + 2.3/b_{ck})(E_{sys}(t) - E_{corr,k}(t))\right]} \quad (4.10)$$

The instantaneous corrosion rate for wire k at time t, $CR_k(t)$ (in mm/yr), can be calculated:

$$CR_k(t) = \frac{(10)(365)(24)(3600) E_w I_k(t)}{96,500d\left\{1 - \exp^{[-(2.3/b_{ak} + 2.3/b_{ck})(E_{sys}(t) - E_{corr,k}(t))]}\right\}} \quad (4.11)$$

If the open-circuit potentials of wires in a WBE are very different from the potential of the entire system (e.g., larger than 100 mV), that is, all wires in a WBE are under a high degree of polarization, equation (4.7) can be applied to estimate localized corrosion using galvanic currents [9], and the instantaneous corrosion rate for wire k at time t, $CR_k(t)$ (in mm/yr), can be calculated using galvanic current data:

$$CR_k(t) = \frac{(10)(365)(24)(3600) E_w I_{gk}(t)}{96,500d} \quad (4.12)$$

However, for a uniform or general corrosion system where $\eta_k = E_{sys} - E_k$ is very small (e.g., approaching zero), I_{ka} calculation from equations (4.6) and (4.9) would give a very large value (e.g., approaching infinite). This suggests that these equations do not apply to a uniform corrosion system.

The kinetics of general corrosion can be determined if the WBE method is employed in conjunction with electrochemical noise resistance (R_n) [13]. The R_n–WBE method [13] is based on the combined use of the R_n method [14–19] and the WBE method [20, 21], allowing R_n measurement from any specific location of a WBE surface and the mapping of instantaneous corrosion rates at all locations on the WBE surface. The noise resistance of a selected wire k (R_{nk}), which is equivalent to the polarization resistance at the location (R_{pk}) [14–19], can be calculated using equation (4.10):

$$R_{nk} = \frac{\sigma V_k(t)}{\sigma I_k(t)} = R_{pk} \quad (4.13)$$

where $\sigma V_k(t)$ is the standard deviation of the potential noise of neighboring wire pair wire k and $k+1$, and $\sigma I_k(t)$ is the standard deviation of the current noise between the same wire pair. $I_k(t)$ is the current noise measured by connecting wire pair k and $k+1$ using a zero-resistance ammeter (ZRA) at time t, and $V_k(t)$ is the potential noise measured simultaneously from the same wire pair. The removal of deterministic low-frequency signals originated from low-frequency electrode processes, such as a slow inhibitor film formation or destruction process, is often necessary to ensure the equivalence between R_{nk} and R_{pk} in equation (4.13) [19]. The trends usually have a nonzero mean value, are not related directly to corrosion processes, and are superimposed on corrosion noise, which is usually a random signal of zero mean value. The simplest form of trend is a linear trend, which has zero frequency and can be removed using the linear method. However, in most practical cases, trends do not follow a specific pattern, and thus it is necessary to have a more general trend removal method. Use of a high-pass filter is a practical method of trend removal. It should be applied to both potential and current noise using the same cutoff frequency (since a low-frequency electrode process should lead to potential and current trends of the same frequency). The noise signals after correct trend removal should have zero mean value—a random signal characteristic. The cutoff frequency should be higher than the frequencies of trends but lower than the dc limit of corresponding impedance spectroscopy measured from the corrosion system [13,19]. The instantaneous corrosion rate for wire k at time t, $CR_k(t)$ (in mm/yr), can be determined using noise resistance measured from over location k, R_{nk} [13]:

$$CR_k(t) = \frac{(10)\,(365)\,(24)\,(3600)\,E_w b_{ak} b_{ck}}{2.3 d\, F\,(b_{ak} + b_{ck})} \frac{1}{R_{nk}} \exp\left\{ \frac{2.3[E_{sys}(t) - E_{corr,k(t)}]}{b_{ak}} \right\}$$

$$(4.14)$$

where $E_{sys}(t)$ is the overall system potential of an integrated WBE system, $E_{corr,k}(t)$ represents the open-circuit corrosion potential of the chosen wire k (disconnected temporarily from the WBE system), and b_{ak} and b_{ck} are, respectively, the Tafel slopes for anodic and cathodic polarization of wire k. E_w is the equivalent weight of electrode material, d is the density of electrode material (7.86 g/cm^3 for steel), and F is the Faraday constant.

Applying equations (4.11), (4.12), or (4.14) to every wire in a WBE, corrosion rates at all locations of the WBE can be calculated based on the overpotential–galvanic current method [9,22], the galvanic current method [22–24], or the R_n–WBE method [25]. A corrosion rate distribution map would give information on instantaneous corrosion rates at time t and also show the patterns of corrosion. Each method has specific advantages, depending on different degrees of electrode uniformity and different electrode noise levels. The question of which equation should be selected for a particular corrosion experiment has been discussed [22].

4.2 MEASURING LOCALIZED CORROSION USING THE OVERPOTENTIAL–GALVANIC CURRENT METHOD

In a typical experiment [9], a mild steel WBE was exposed to several corrosive media containing 0.017 M sodium chloride and 0.008 M sodium carbonate or 0.5 M

H_2SO_4 under static partial or total immersion conditions. The potential and galvanic current distributions over the electrode surfaces were measured and visual observations were made at different stages of the exposure tests. During exposure periods, all the wire terminals of a WBE were connected together so that electrons would move freely among wires similar to the way they would for mild steel electrodes and plates with a larger surface area. An Ag/AgCl reference electrode was used for corrosion potential measurements. An automatic zero-resistance ammeter (AutoZRA, ACM Instruments, England) was used to record the corrosion potential and galvanic current. The AutoZRA enables current (325 mA to 10 pA) and voltage to be measured accurately and to be recorded automatically. Its data logging software runs in a Microsoft Windows environment, complete with a real-time Excel link. Thus, the potential and current data analysis and plotting can be performed using Microsoft Excel. In this experiment, the corrosion potential of the multiminielectrode system (E_{sys}) and of each wire (E_k for wire k), the galvanic currents between wires and the system (I_{gk} for wire k), and the Tafel slopes (b_{ak} and b_{ck}) were measured experimentally.

When measuring the corrosion potentials of wires (E_k for wire k) against an Ag/AgCl reference electrode, the terminals of wires in the electrode were connected manually, in sequence, to the AutoZRA. There was a delay of 10 seconds between connecting a wire to the AutoZRA and making the measurement. The experimental design for measuring the corrosion potentials of an individual wire (total immersion) is shown in Figure 4.3. The measurements were repeated regularly during the experimental period. The measurements of galvanic currents (I_{gk} for wire k) flowing between each wire and the system were carried out by using the experimental design shown in Figure 4.4. An AutoZRA was connected manually in sequence with a particular wire terminal and with the other terminals. The system corrosion potential (E_{sys}) was also measured using this experimental design by employing a voltmeter against a reference electrode. A typical experimental design for measuring the Tafel slopes by performing a linear polarization measurement

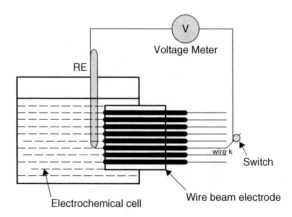

Figure 4.3 Corrosion potential measurement using a WBE (total immersion). (From [9].)

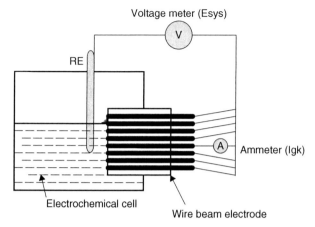

Figure 4.4   Galvanic current and the system corrosion potential measurements using a WBE (partial immersion). (From [9].)

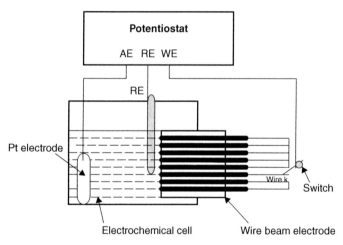

Figure 4.5   Measurement of Tafel slopes (b_{ak} and b_{ck} for wire k) of individual wires (total immersion). (From [9].)

or by performing Tafel curve measurement on each wire is shown in Figure 4.5. Measurements were repeated regularly during the experiment. An EG&G Potentiostat/Galvanostat Model 273A (Princeton Applied Research) was used for this purpose.

Waterline corrosion is a well-known localized corrosion phenomenon [8]. When a steel surface is partly immersed in an unstirred solution containing a corrosive salt (e.g., sodium chloride) with an insufficient amount of inhibitor (e.g., sodium carbonate), intense localized corrosion occurs along the waterline areas on the plate. In the present experiment [9], similar localized corrosion behavior was observed along the waterline area on the mild steel plate and also on the surface of the

WBE, which had been partially immersed in a 0.017 M NaCl and 0.008 M Na_2CO_3 solution. Figure 4.6 shows results from the measurements of potential and galvanic current distributions and visual observations over the WBE surface.

Figure 4.6 shows the gradual initiation and propagation of a localized corrosion pattern over the WBE surface. Initially, as shown in Figure 4.6a, there were only a few isolated anodic sites, indicated by the positive galvanic currents, among a large number of cathodic sites. Brown corrosion product dots were observed only at those isolated anodic sites. In later stages, as shown in Figure 4.6b, c, and d, some of the anodic sites merged, forming larger localized corrosion zones, while other anodic sites stopped growing. There is very good agreement between visual observations and the corrosion potential and galvanic current distribution [9]. Figure 4.6 clearly illustrates localized corrosion due to a large cathodic zone and small anodic zones. The significant potential difference between cathodic and anodic areas (maximum 400 mV) produced large galvanic currents and resulted in localized corrosion. Corrosion was concentrated on anodic areas, and the large cathodic areas were protected.

Very different corrosion patterns were observed when the WBE and a steel plate were partially immersed in an inhibitor-free 0.017 M NaCl solution. A general corrosion attack occurred on almost all the surface areas of the WBE and mild steel plate except at the waterline area. Figure 4.7 shows results of potential and galvanic current measurements and visual observations at various stages of the exposure test. Initially, as shown in Figure 4.7a, localized corrosion was concentrated on areas close to the waterline. It was a small anode and large cathode corrosion system. After 18 hours' exposure, as shown in Figure 4.7b, the anodic areas expanded. Finally, as shown in Figure 4.7c, the majority of the electrode surface became anode. As a result, the electrode surface became a large anode and small cathode corrosion system. The areas just below the waterline behaved as a cathode and remained unattacked. Good agreement between visual observations and the corrosion potential and galvanic current distribution was again obtained [9].

Although this type of corrosion is usually referred to as general corrosion, it is actually localized corrosion in nature. There were large potential differences between different areas of the electrode surface. Figures 4.6 and 4.7 show clearly that the surface of a WBE can effectively simulate the corrosion processes occurring on a large-area electrode–plate surface. In the system studied, the size of wire did not limit the progress of localized corrosion. This work also demonstrated that WBEs can be used to monitor and study localized corrosion processes. It has been shown that WBEs can relate the anodic process of corrosion to the anodic regions of the corroding surface and the cathodic process to the cathodic regions–and so to study separately the anodic and cathodic processes of localized corrosion cells.

Steel corrosion in 0.5 M H_2SO_4 has traditionally been treated as typical uniform corrosion, and traditional electrochemical techniques have commonly been applied to such systems. Indeed, during the 28 hours' total exposure test, both surfaces of the WBE and mild steel plate were visually clean and uniform [9]. However, as shown in Figure 4.8a, the potential distribution over the WBE surface immersed in 500 mL of 0.5 M H_2SO_4 for 1 hour was actually not uniform. There was a

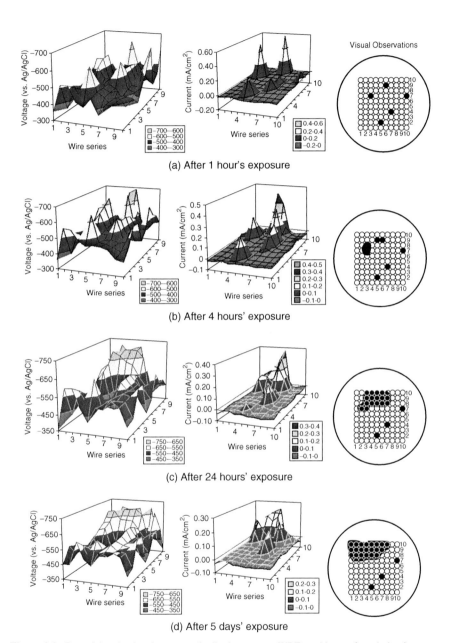

(a) After 1 hour's exposure

(b) After 4 hours' exposure

(c) After 24 hours' exposure

(d) After 5 days' exposure

Figure 4.6   Potential and galvanic current distributions over a WBE working surface during its exposure to $0.017\,M$ NaCl $+ 0.008$ M Na_2CO_3 solution. (From [9].) (*See insert for color representation of the figure.*)

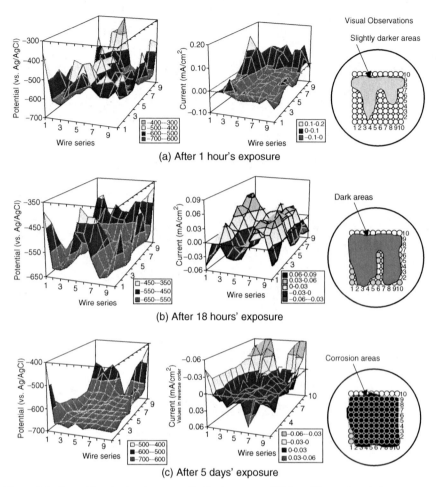

Figure 4.7 Potential and galvanic current distributions over a WBE working surface during exposure to 0.017 M NaCl solution. (From [9].) (*See insert for color representation of the figure.*)

maximum 30 mV potential difference between different locations over the WBE surface. This suggests that a steel electrode surface in 0.5 M H_2SO_4 is, in fact, an electrochemically nonuniform corrosion system. The reason that electrode surfaces in 0.5 M H_2SO_4 are visually uniform is shown in Figure 4.8, which highlights the very important characteristic that the cathodic and anodic sites of corrosion were generally distributed in a random manner and changed with time. For example, the cathodic and anodic sites in Figure 4.8d and f are different from those in Figure 4.8b. This random feature and variability of corrosion sites can result in a uniform electrode surface. Unlike localized corrosion, in the case of visually uniform corrosion, anodic and cathodic areas are generally randomly distributed and change with time. This observation is in agreement with the uniform corrosion models described in Chapter 1.

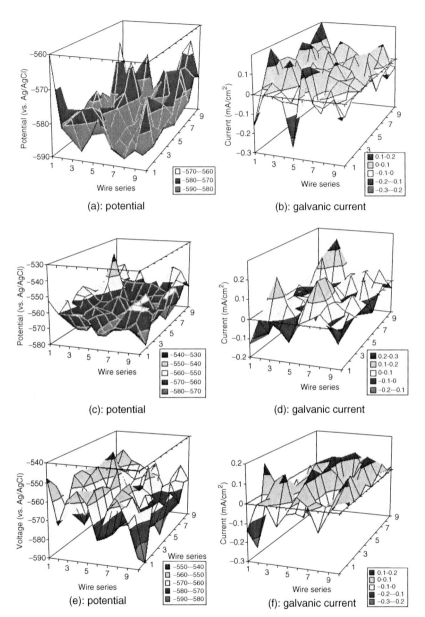

Figure 4.8   Potential and galvanic current distributions over a WBE surface after being exposed to 0.5 M H₂SO₄ solution. (From [9].) (*See insert for color representation of the figure.*)

Instantaneous localized corrosion currents at various stages of the exposure tests could be estimated using equation (4.11) or (4.12). Equation (4.11) is used if the voltage difference between localized anodes and cathodes is small, whereas equation (4.12) is used if there is a large voltage difference between the open-circuit potential of a wire and the potential of the entire system (e.g., >100 mV). To test the accuracy of the WBE measurements, a reference technique is required. For the purpose of comparison, in this work instantaneous localized corrosion currents, measured using equation (4.11) or (4.12), were used to calculate average localized corrosion currents over the entire exposure period and then to estimate average localized corrosion rates using Faraday's law [9]. The average localized corrosion rates can be compared with corresponding results from solution analysis. The iron concentration in the solutions employed was measured using atomic absorption analysis. After exposure tests, corrosion products were removed from the electrode surfaces, dissolved in HCl, and added to the solutions. The iron dissolved in the solution is related directly to the average localized corrosion rates at anode zones because the dissolved iron came mainly from the corrosion reaction at anode zones. The corrosion in cathodic zones can be ignored because the cathode areas of the electrode–plates were visually clean and free of corrosion products. A comparison between this new method and traditional solution analysis shows good correlation (Table 4.1).

This experiment demonstrates that the working surface of a WBE effectively simulates a large-area electrode–plate surface in its corrosion behavior. The size of wire did not limit the progress of localized corrosion. A WBE can relate the anodic process of corrosion to the anodic regions of the corroding surface and the cathodic process to the cathodic regions and thus make it possible to study the anodic and cathodic processes of the localized corrosion cells separately. The WBE method is able to measure corrosion rates and to monitor localized corrosion processes and patterns.

Experimental findings from this typical experiment were verified further in a longer-term experiment [12]. Figure 4.9 shows corrosion potential distribution maps measured from a WBE surface immersed in a corrosive electrolyte containing 0.017 M sodium chloride and 0.008 M sodium carbonate using the experimental

Table 4.1 Comparison of Localized Corrosion Rates Obtained Using the WBE Method and Solution Analysis

Experimental Conditions	Corrosion Pattern	Corrosion Rate (mm/yr)	
		WBE Method	Solution Analysis
Partial exposure to 0.017 M NaCl + 0.008 M Na$_2$CO$_3$ solution for about 6 days	Waterline corrosion and localized attack	1.39	1.32
Partial exposure to 0.017 M NaCl solution for about 5 days	Localized corrosion	0.34	0.37
Total exposure to 0.017 M NaCl solution for about 20 hours	Localized corrosion	0.13	0.19

Source: [9].

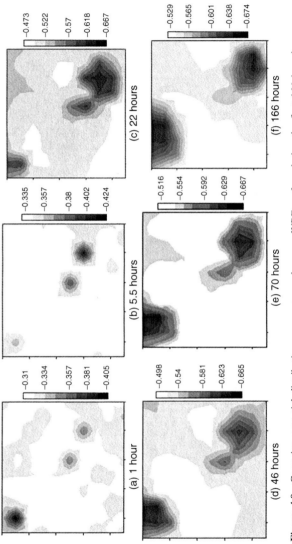

Figure 4.9 Corrosion potential distribution maps measured over a WBE surface during the first 166 hours' exposure (potential in volts). (From [12].)

117

setup shown in Figures 4.3 to 4.5 and the experimental method detailed above. It is clear that the WBE surface was thermodynamically heterogeneous, indicated by the nonuniform potential distribution, and that this heterogeneity was propagated with the extension of electrode immersion. At the initiation stage (Figure 4.9a and b), there were only a few isolated anodic sites, indicated by the relatively low corrosion potential values, among a large number of cathodic sites. This electrochemical heterogeneity propagated at later stages, as shown in Figure 4.9c to f. Some of the anodic sites gradually merged, forming larger anodic zones, while other anodic sites stopped growing, and this propagation process was not limited by the insulating layers between neighboring wires. It is expected that the WBE surface would reach a steady state, and miniequilibria would be established over various locations on the WBE surface.

Experimentally measured corrosion potentials (Figure 4.9), corresponding galvanic currents, and Tafel constants were employed to determine electrochemical corrosion kinetics using equation (4.11). Maps showing instantaneous corrosion dissolution rates over a WBE surface at a particular point in immersion time were obtained (Figure 4.10). Figures 4.9 and 4.10 show clearly that the WBE surface was thermodynamically and kinetically heterogeneous. The kinetic heterogeneity and its propagation processes (Figure 4.10) correlated well with the thermodynamic heterogeneity and its propagation (Figure 4.9). This result suggests that the thermodynamic difference over the electrode surface, indicated by $E_{sys} - E_{corr,k}$ in equation (4.11), resulted in kinetic heterogeneity. This heterogeneity obviously led to the accumulation and localization of surface chemistry changes, which in return affected the thermodynamic heterogeneity over the electrode surface, leading to a distinct separation of electrode surface chemistry and reactions. The propagation of electrochemical heterogeneity appeared to be a self-catalytic process—a phenomenon that had previously been observed in localized corrosion such as pitting corrosion [26]. In this process, electrons traveled continuously from the anodic to the cathodic areas through the electrode body, and at the same time, ions traveled between anodic and cathodic areas through the electrolyte, resulting in rapid localized corrosion penetration at the anodic areas. This result also suggests that the origin of electrochemical heterogeneity is the thermodynamic heterogeneity in an electrochemical system. Careful electrode surface preparation and electrolyte stirring could help alleviate this thermodynamic heterogeneity; however, they could not eliminate it since a practical electrode, regardless of its shape and size, is heterogeneous in nature.

Instantaneous corrosion rate (CR) maps such as those shown in Figure 4.10 can be used to determine total corrosion depths over an entire experimental period. This was achieved by summing up the average corrosion depths calculated, which corresponds to the instantaneous corrosion rates measured after various periods of exposure, to give a cumulative result. After completing the experiment, the surface of the electrode was cleaned in an acid solution containing inhibitor, which was made by adding 20 g of Sb_2O_3 and 59 g of $SnCl_2 \cdot 2H_2O$ into 1 L of concentrated HCl for 1 minute to remove corrosion products. Then the surface of the electrode was carefully observed under a microscope to determine the depth of corrosion

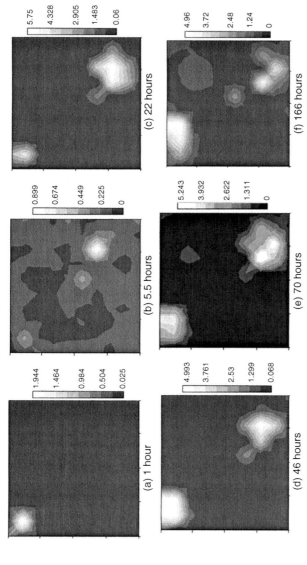

Figure 4.10 Corrosion rate distribution maps measured over a WBE surface during the first 166 hours' exposure (corrosion rates in mm/yr). (From [12].)

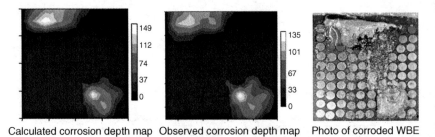

Calculated corrosion depth map   Observed corrosion depth map   Photo of corroded WBE

Figure 4.11 Corrosion depth maps calculated and observed microscopically and a photograph of the corroded surface after 406 hours' exposure (corrosion depths in µm) [12]. (*See insert for color representation of the figure.*)

and its distribution over the electrode surface. This was done using a microscope (Olympus Model PMG3) by focusing on each corroded wire surface in turn and then on the adjacent insulating epoxy layer surface. The difference in focusing distance was used to estimate the depth of corrosion for the wire selected. Figure 4.11 shows corrosion depth distribution maps calculated and observed microscopically and a photograph of the corroded WBE after being exposed to a corrosive environment for 406 hours. Indeed, the total corrosion depth calculated correlates quantitatively with the corrosion depths observed microscopically and the photograph of the corroded WBE. This result confirms the accuracy of the thermodynamic and kinetic measurement methods.

4.3 MEASURING LOCALIZED CORROSION USING THE GALVANIC CURRENT METHOD

Equation (4.12) is a more convenient method of calculating localized corrosion kinetics if there is a large voltage difference between the open-circuit potential of a wire and the potential of the entire system (e.g., > 100 mV). In a typical experiment [24] this method was used to measure waterline corrosion of steel over a long exposure period of 404 days to achieve a better understanding of practical long-term corrosion processes and kinetics. In this experiment, a freshly polished WBE surface was exposed to a solution containing 0.017 M NaCl and 0.008 M Na_2CO_3 under partial immersion static conditions to allow waterline corrosion to occur (Figure 4.4). During exposure, all the wire terminals of a WBE were connected together so that electrons could move freely between wires, similar to that which would be the case with mild steel electrodes–plates with a larger surface area. An Ag/AgCl reference electrode was used for corrosion potential measurement. Corrosion rates were calculated based on galvanic current data only because galvanic current can be used to estimate instantaneous localized corrosion rates at the anodic zone whenever there is a large potential difference between the anodic and cathodic zones. This is obviously valid in this experiment since during the exposure period, there is always a very large potential difference between the

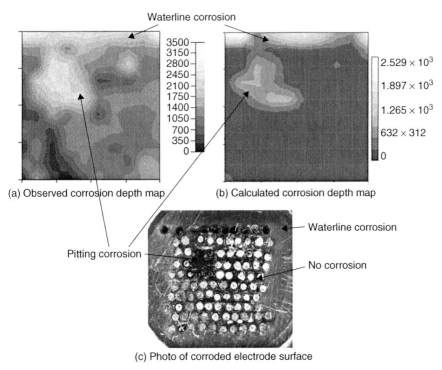

(a) Observed corrosion depth map   (b) Calculated corrosion depth map

(c) Photo of corroded electrode surface

Figure 4.12   Corrosion depth maps and values (in μm) observed and calculated, plus a photograph showing a WBE probe surface after exposure to a waterline environment for 404 days. (*See insert for color representation of the figure.*)

anodic and cathodic zones (typically, 150 mV). The anode zones remained largely fixed during the exposure period. During 404 days' immersion, localized corrosion occurred on the WBE surface (Figure 4.12). Serious corrosion occurred mainly along the waterline area and a pitting area just under the waterline. The corrosion depths shown in Figure 4.12 were calculated by accumulating instantaneous corrosion penetrations over the entire test duration. Figure 4.13 shows instantaneous corrosion rate maps estimated using galvanic current data, measured over various periods of exposure. These maps clearly indicate that localized corrosion occurred over the electrode surface at varied rates. The corrosion centers also changed over the exposure period. There is a clear correlation between the corrosion depth maps calculated and observed and the photograph of the corroded WBE surface. The corrosion depth maps in Figure 4.12 give similar, although not identical, corrosion depth values.

After the WBE surface was exposed to corrosive media for 1 hour, as shown in Figure 4.13a, electrochemical heterogeneity developed over the electrode surface, and localized corrosion occurred. With the extension of exposure (Figure 4.13b and c, some of the corrosion sites in Figure 4.13a stopped and corrosion was concentrated in a small area just below the waterline. At this stage, the majority

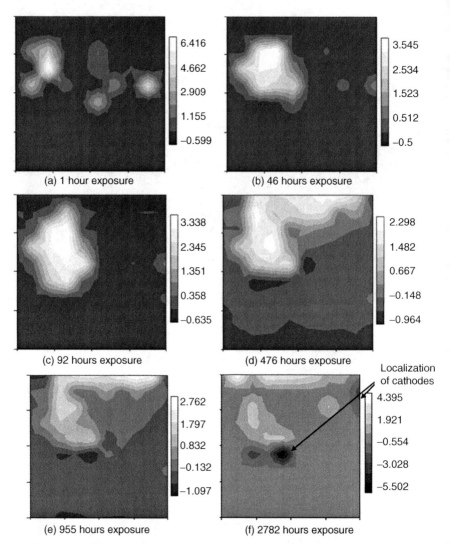

Figure 4.13 Corrosion rate values (in mm/yr) over a WBE after exposure to a waterline corrosion environment for various durations during the 404-day exposure experiment.

of the WBE surface area behaved as cathodes. The negative corrosion rates shown in the maps were calculated from cathodic currents according to Faraday's law (here the cathodic process is the reduction of oxygen). They indicate only cathodic reaction activities and do not imply that the metal is getting thicker. Usually, these negative corrosion rates are replaced by zero corrosion rates since no metal dissolution or deposition occurred over cathodic areas. These negative corrosion rates on corrosion maps clearly indicate cathodic reaction activities and cathode

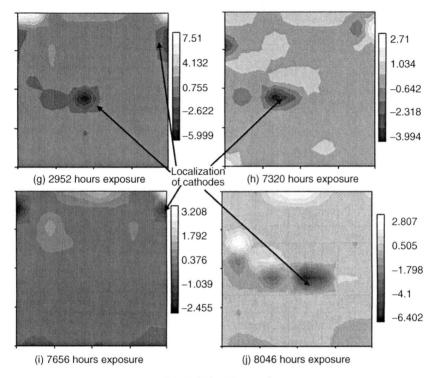

Figure 4.13 (*Continued*)

locations. With further exposure, as shown in Figure 4.13d to f, waterline corrosion began and propagated. Corrosion became highly localized and, interestingly, there was also an obvious localization of cathodes (negative corrosion rate sites). Over a very long period, as shown in Figure 4.13g to j, corrosion concentrated primarily on the waterline area, and localization of cathodes remained. Higher corrosion rates were also experienced in the pitting area. As an accumulated result, the electrode surface was locally corroded, with the appearance of both waterline and pitting corrosion (Figure 4.12).

In a highly localized corrosion system, corrosion rate distribution maps calculated using the overpotential–galvanic current method and the galvanic current–only method are often very similar and correlate well with corrosion characteristics in the system. Figure 4.14 shows corrosion potential and corrosion rate maps measured after a WBE that was exposed to a stagnant crevice corrosion environment for 3 hours [23]. The two methods generated very similar maps because this system had significant potential differences between wires (maximum nearly 300 mV), and thus galvanic current can be used to estimate instantaneous localized corrosion rates in the anodic zone.

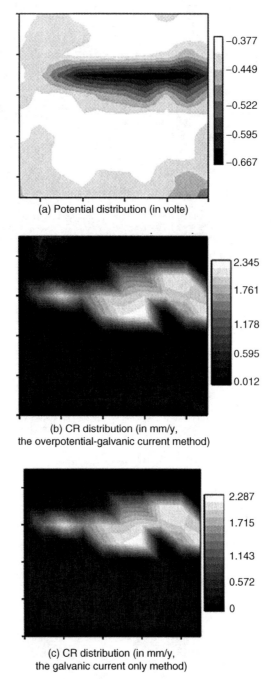

(a) Potential distribution (in volte)

(b) CR distribution (in mm/y,
the overpotential-galvanic current method)

(c) CR distribution (in mm/y,
the galvanic current only method)

Figure 4.14 Corrosion potential and corrosion rate measured after a WBE was exposed to a stagnant oxygen crevice corrosion environment for 3 hours. (From [23].)

4.4   MEASURING LOCALIZED CORROSION USING THE R_n–WBE METHOD

Equations (4.6) and (4.9) do not apply to uniform corrosion systems where $\eta_k = E_{sys} - E_k$ is very small (e.g., approaching zero). The R_n–WBE method has been proposed to overcome this problem [13]. In this method, a WBE is used in conjunction with the electrochemical noise resistance to determine the kinetics of relatively uniform corrosion systems. Figure 4.15 shows a typical experiment in which corrosion of mild steel in a carbon dioxide–saturated brine is measured using the R_n–WBE method. To establish different degrees of nonuniformity across the surface of the WBE, the working surface was filmed with an inhibitor layer by placing a drop of inhibitor on the WBE surface. The electrochemical cell was charged with 800 mL of synthetic brine. Two types of corrosion conditions were employed to allow different forms of CO_2 corrosion to occur. The corrosion cell shown in Figure 4.15 created a localized CO_2 corrosion environment since a CO_2 sparging tube was arranged to generate a turbulent flow over a portion of the electrode surface. Carbon dioxide was sparged directly onto the electrode surface, and this resulted in a turbulent flow that damaged the inhibitor film on the electrode

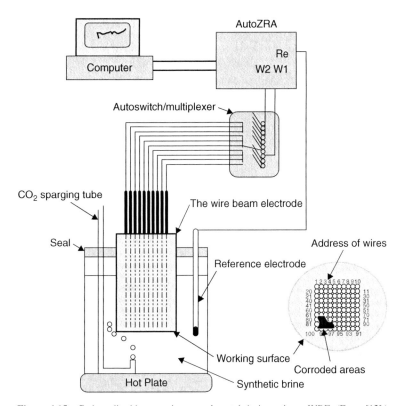

Figure 4.15   Carbon dioxide corrosion experimental design using a WBE. (From [13].)

surface locally and caused a localized form of corrosion. Alternatively, the sparging tube can be rearranged to keep CO_2 bubbles away from the WBE working surface. Under these conditions, the inhibitor film on the electrode surface deteriorates in a general way, which results in a general form of corrosion. The temperature was kept at $70°C$ [13].

The electrochemical noise resistance method was used in this experiment to measure the polarization resistance from various locations of a corroding WBE surface since the noise resistance method is technically more convenient than are conventional methods of polarization resistance measurement [13–19]. For example, to measure noise resistance there is no need to apply perturbation to the test system by externally imposed polarization, and the instrument system is simple. The measurement of noise resistance distribution over a WBE surface requires the measurement of current noise and potential noise from the WBE system. An AutoZRA was connected in sequence to neighboring pairs of wires in a WBE using a preprogrammed automatic switch device. The measurement was carried out by connecting the AutoZRA to wires 1 and 2, then wires 2 and 3, wires 3 and 4, . . ., wires k and $k+1$, . . ., wires 99 and 100. The current noise is the galvanic current fluctuation between these wire pairs. The potential noise is the potential fluctuation of these wire pairs (connected together) against an Ag/AgCl reference electrode. The noise resistance R_{nk} of a selected wire k can be calculated with equation (4.13) using the standard deviation of the potential and current noise, $\sigma V_k(t)$ and $\sigma I_k(t)$, of the neighboring wire pair wire k and wire $k+1$. In this application, the neighboring wires are assumed to be identical working electrodes, an experimental requirement for noise resistance measurement. This assumption is reasonable since wires in a WBE have a very small surface area and are adjacent to each other; thus, neighboring wires are very likely under similar electrochemical conditions. R_{nk} is measured from every local area of the WBE surface and is plotted to produce a noise resistance distribution map.

Experimentally measured corrosion potential distribution, galvanic current distribution, noise resistance distribution and Tafel constants over a WBE surface after various periods of electrode exposure were used to calculate corrosion rates and their distribution over the electrode surface using equation (4.14). This corrosion rate distribution thus gives the instantaneous corrosion dissolution rate of the metal at a particular point in time. Total corrosion depths over the entire experimental period can then be calculated from these corrosion rate maps measured after different exposure periods. This was achieved by summing the average corrosion depths calculated corresponding to the instantaneous corrosion rates measured after various periods of exposure to give a cumulative result. For wire k,

$$\text{TotalDepth}_k = \text{CorrDepth}(1)_k + \text{CorrDepth}(2)_k + \text{CorrDepth}(3)_k$$

$$+ \cdots + \text{CorrDepth}(i)_k + \cdots \qquad (4.15)$$

where $\text{CorrDepth}(i)_k$ is the incremental corrosion depth that results during a specific exposure period i for wire k. The corrosion depth at any period selected can be calculated using the equation

$$\text{CorrDepth}(i)_k = \text{CR}_k \cdot \frac{1000 \cdot \text{Period}}{(24)(365)(3600)} \tag{4.16}$$

where $\text{CorrDepth}(i)_k$ is the corrosion depth of the wire k (in μm) during a specific period i. CR_k is the corrosion rate of wire k over period i, which normally is 24 hours. During the 24-hour period, the corrosion rate is assumed to be constant. Quantitatively, TotalDepth_k should correlate closely with the real corrosion depth of wire k, which is measurable by microscope at the end of the total exposure period, given the approximation that the corrosion rate remained constant between actual measurements. In this way, a corrosion depth distribution map can be produced.

A WBE was exposed to a localized corrosion environment (Figure 4.15) with a CO_2 sparging tube kept close to wire 91 of the electrode surface. Carbon dioxide was sparged directly onto the electrode surface, which resulted in a turbulent flow that could locally damage the inhibitor film on the electrode surface and cause localized corrosion. Indeed, as shown in Figure 4.16, the electrode surface clearly exhibited localized corrosion. Areas around wire 91, close to the CO_2 sparging

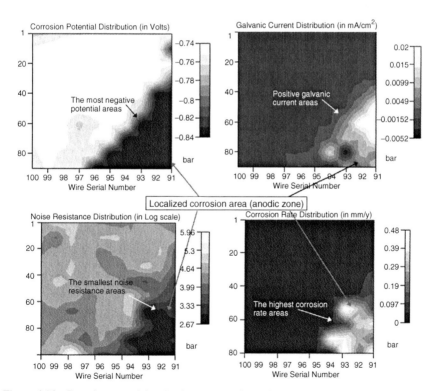

Figure 4.16   Corrosion potential, galvanic current, noise resistance, and corrosion rate distribution maps measured from a WBE after exposure to a simulated CO_2 corrosion cell.

tube, behaved as anodes and were subject to rapid corrosion dissolution. This is clearly indicated by the more negative values of corrosion potential and positive values in galvanic current (indicating electrons leaving the area) recorded in this zone. The potential distribution map in Figure 4.16 shows a nonuniform potential distribution and there is a maximum potential difference of more than 100 mV over the electrode surface. The most negative corrosion potential values were recorded around the area where wire 91 was located. This appears to have arisen from the turbulent flow, which damaged the inhibitor film in that area. An unprotected metallic surface would obviously form the anode of the corrosion system. In the noise resistance distribution map shown in Figure 4.16, noise resistance values in the anodic areas were much lower than in other areas, resulting in a pattern similar to that of other maps in Figure 4.16. The clear correlation between the noise resistance map and the other maps suggests that in the anodic areas around wire 91, not only was galvanic corrosion occurring, but uniform corrosion was also occurring with a significantly higher corrosion rate than in other areas. This suggests that both uniform and galvanic corrosion components contribute to the localization of corrosion. It also confirms that the inhibitor film on the areas around wire 91 was indeed locally damaged by turbulent flow and became less protective. The potential, galvanic current, and noise resistance distribution data were used to calculate corrosion rate distribution over the corroding electrode surface. Indeed, as shown in the corrosion rate distribution map of Figure 4.16, corrosion rates on the anodic areas were significantly higher than those in other areas. Overall, there is a good correlation between all the maps in the figure.

To confirm the accuracy of the corrosion rate distribution maps, total accumulative corrosion depths and the distribution of these depths were calculated from corrosion rate maps. Obviously, if corrosion rate distribution maps such as those in Figure 4.16 are accurate, this total corrosion depth distribution map should correlate quantitatively with the real corrosion depth and its distribution, which is measurable with a microscope at the end of the total exposure period. As shown in Figure 4.17, the corrosion depth values calculated are similar, although not identical, to the values measured microscopically.

Another WBE was exposed to another carbon dioxide corrosion cell with a corrosion environment that was slightly different from the corrosion environment used previously in the work described above. The CO_2 sparging tube in the corrosion cell was kept away from the probe surface, and thus carbon dioxide was not sparged directly onto the electrode surface. As a result, there was no turbulent flow on the electrode surface, so the corrosion conditions over the electrode surface were relatively uniform. Indeed, after 408 hours' exposure, corrosion of the electrode surface was less localized and had the appearance often referred to as general corrosion. Compared to the maps shown in Figure 4.16, maps obtained from this general corrosion system show different characteristics. A typical result is shown in Figure 4.18. In this system electrochemical parameters recorded from different locations of the electrode surface were similar; for example, the maximum potential difference between the anodic and cathodic zones was only about 30 mV. Furthermore, across the entire duration of the test, the majority of the electrode

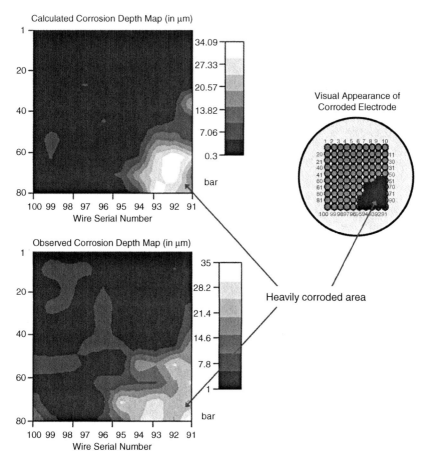

Figure 4.17   Corrosion depth maps and values (in μm) observed and calculated over a WBE surface after exposure to a simulated CO_2 corrosion environment for 241 hours.

surface was undergoing corrosion, typically as shown in Figure 4.18. Corrosion rate values recorded from various locations of the electrode surface were relatively close. The range of corrosion rates observed across the entire probe surface was also narrower. These observations are consistent with the fact that the electrode surface exhibited less localization in its corrosion pattern. In general, the corrosion potential, galvanic current, noise resistance, and corrosion rate distribution maps in Figure 4.18 correlate well.

To confirm the accuracy of corrosion rate distribution maps such as those shown in Figure 4.18, accumulative corrosion depths and their distribution were calculated from corrosion rate maps obtained at various stages of exposure. A corroded electrode surface was observed and measured under a microscope at the end of the experiment to obtain an observed corrosion depth map. Figure 4.19 shows observed and calculated corrosion depth maps that have generally similar features and

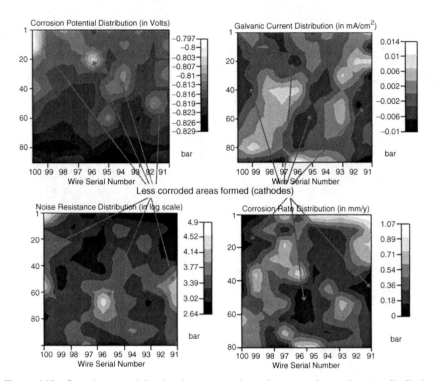

Figure 4.18   Corrosion potential, galvanic current, noise resistance, and corrosion rate distribution maps measured from a WBE after exposure to a CO_2 corrosion cell.

corrosion depth values. This result demonstrates that the electrochemical method also works well even if the corrosion is not particularly localized. In principle, the technique described should apply effectively to any form of corrosion process.

It should be pointed out that although the corrosion depth maps in Figures 4.17 and 4.19 have similar features and corrosion depth values, the correlations are not perfect. Several factors may have contributed to this. First, equation (4.14), which was used to calculate the corrosion rate, is based on several assumptions that a real corrosion system may not fully obey. Second, corrosion depth calculations according to equations (4.15) and (4.16) was based on a limited number of corrosion rate measurements (usually, one measurement every 24-hours) and the corrosion rate during this 24-hour period was assumed to be constant. However, in reality, the corrosion rate may vary during this period. Third, the corrosion depth measurement using a microscope is not perfectly accurate, and there may therefore be some uncertainty in the observed corrosion depth map. Another source of error is the possible inaccuracy of noise resistance measurement. In this experiment, the R_n measurement of each pair of wires had a relatively short duration of 10 seconds, with a sampling rate of 30 points/second (i.e., the frequency range of this measurement extended from 15 Hz to 0.1 Hz). This frequency range does indeed cover the dc limit of the corresponding impedance spectrum [13], and thus the R_n

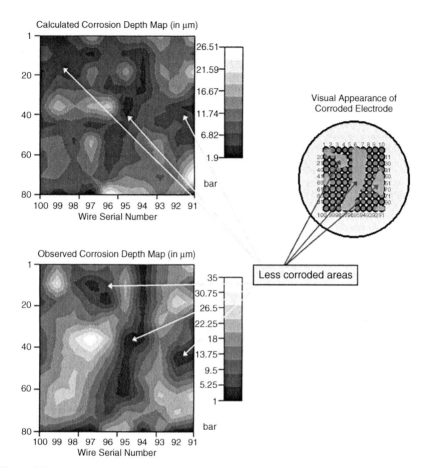

Figure 4.19 Corrosion depth maps and values (in μm) observed and calculated over a WBE surface after exposure to a simulated CO_2 corrosion environment for 408 hours.

values measured are acceptable. However, care should be taken when this method is applied to systems where the dc limit is lower than 0.1 Hz. The applicability of the R_n–WBE method has been found to depend on the noise level of a corroding electrode surface and was found not applicable to low-noise-level corrosion systems [13].

REFERENCES

1. ASTM G61-86, *Standard Test Method for Conducting Cyclic Potentiodynamic Polarization Measurements for Pitting Corrosion Susceptibility of Iron-, Nickel-, or Cobalt-Based Alloys*, ASTM, West Conshohocken, PA, 2009.
2. G. T. Burstein, C. Liu, R. M. Souto, and S. P. Vines, The origins of pitting corrosion, *Corrosion Engineering, Science, and Technology*, 39 (2004), 25–30.

3. P. Schmutz and G. S. Frankel, Characterization of AA2024-T3 by scanning Kelvin probe force microscopy, *Journal of the Electrochemical Society*, 145 (1998), 2285–2295.

4. J. H. W. de Wit, Local potential measurements with the SKPFM on aluminum alloys, *Electrochimica Acta*, 49 (2004), 2841–2850.

5. G. H. Kelsall, in *Techniques in Electrochemistry, Corrosion and Metal Finishing*, A. T. Kuhn, Ed., Wiley, New York, 1987.

6. A. J. Bard and L. R. Faulkner, *Electrochemical Methods: Fundamentals and Application*, 2nd ed., Wiley, New York, 2000.

7. F. Mansfeld, Classic paper in corrosion science and engineering with a perspective by F. Mansfeld, *Corrosion*, 62 (2006), 843.

8. U. R. Evans, *An Introduction to Meallic Corrosion*, 3rd ed., Edward Arnold, London, 1981.

9. Y. J. Tan, Monitoring localized corrosion processes and estimating localized corrosion rates using a wire-beam electrode, *Corrosion*, 54 (1998), 403.

10. J. A. V. Butler, Studies in heterogeneous equilibria: I and II. Conditions at the boundary surface of crystalline solids and liquids, and the application of statistical mechanics; and The kinetic interpretation of the Nernst theory of electromotive force, *Transactions of the Faraday Society*, 19 (1924), 729, 734.

11. M. Volmer and T. Erdey-Gruz, The theory of hydrogen overvoltage, *Zeitschrift fuer Physikalische Chemie*, 150A (1930), 203.

12. Y. J. Tan, Corrosion science: a retrospective and current status, presented at the Electrochemical Society 201st Meeting, Philadelphia, G. S. Frankel, H. S. Isaacs, J. R. Scully, and J. D. Sinclair, Eds., 2002, PV2002-13.

13. Y. J. Tan, S. Bailey, B. Kinsella, and A. Lowe, Mapping corrosion kinetics using the wire beam electrode in conjunction with electrochemical noise resistance measurements, *Journal of the Electrochemical Society*, 147 (2000), 530–540.

14. D. A. Eden, K. Hladky, D. G. John, and J. L. Dawson, Electrochemical noise resistance, Paper 274, presented at Corrosion' 86, NACE, Houston, TX, 1986.

15. F. Mansfeld and H. Xiao, Electrochemical noise analysis of iron exposed to NaCl solutions of different corrosivity, *Journal of the Electrochemical Society*, 140 (1993), 2205.

16. U. Bertocci, C. Gobrielli, F. Huet, and M. Keddam, Noise resistance applied to corrosion measurements: I. Theoretical analysis, *Journal of the Electrochemical Society*, 144 (1997), 31–37.

17. R. A. Cottis, S. Turgoose, and R. Newman, *Corrosion Testing Made Easy: Electrochemical Impedance and Noise*, NACE International, Houston, TX, 1999.

18. R. A. Cottis, Interpretation of electrochemical noise data, *Corrosion*, 57 (2001), 265–285.

19. Y. J. Tan, S. Bailey, and B. Kinsella, The monitoring of the formation and destruction of corrosion inhibitor films using electrochemical noise analysis, *Corrosion Science*, 38 (1996), 1681.

20. Y. J. Tan, Sensing electrode inhomogeneity and electrochemical heterogeneity using an electrochemically integrated multi-electrode array, *Journal of the Electrochemical Society*, 156 (2009), C195–C208.

21. Y. J. Tan, Sensing pitting corrosion by means of electrochemical noise detection and analysis, *Sensors and Actuators B*, 139 (2009), 688–698.

22. Y. J. Tan, An experimental comparison of three wire beam electrode based methods for determining corrosion rates and patterns, *Corrosion Science*, 47 (2005), 1653–1665.

23. Y. J. Tan, S. Bailey, and B. Kinsella, Mapping non-uniform corrosion in practical corrosive environments using the wire beam electrode method: II. Crevice corrosion, *Corrosion Science*, 43 (2001), 1919–1929.

24. Y. J. Tan, S. Bailey, and B. Kinsella, Mapping non-uniform corrosion in practical corrosive environments using the wire beam electrode method: III. Water-line corrosion, *Corrosion Science*, 43 (2001), 1930–1937.

25. Y. J. Tan, S. Bailey, and B. Kinsella, Mapping non-uniform corrosion in practical corrosive environments using the wire beam electrode method: I. Multi-phase corrosion, *Corrosion Science*, 43 (2001), 1905–1918.

26. G. Fontana, *Corrosion Engineering*, 3rd ed., McGraw-Hill, New York, 1987.

5

Characterizing Inhomogeneity and Localized Corrosion on Coated Electrode Surfaces

An electrode surface covered by an organic coating or an inhibitor film is likely to be more inhomogeneous than are bare electrode surfaces because a coated electrode surface could be affected not only by preexisting inhomogeneities on the electrode substrate, such as oxides, rust, mill scale, dust, grease, oil, salts, and old coatings, but also by inhomogeneities existing in the coating or film, such as air bubbles, cracks, microvoids, contaminants, trapped solvents, and nonbonded and weak areas. Inhomogeneities on coated electrode surfaces have long been known to affect underfilm corrosion significantly. A common experience is that rust on a coated metal surface is normally initiated at localized weaker areas of the coating.

Inhomogeneities on the substrate surface are believed to behave as the initiation sites of coating failure and to influence underfilm corrosion by affecting the coating adhesion [1]. Inhomogeneities in a coating film are believed to influence the anticorrosion performance of a coating system significantly by affecting factors such as the permeability and the transportation of aggressive species such as water, oxygen, and cations through the coating and along the coating–substrate interface [2]. For example, a nonuniform cross-link in a coating film was found to result in local D (direct) and I (indirect) sites that lead to major variations in coating resistances and anticorrosion behavior [3]. Pigments in a coating film could lead to local

Heterogeneous Electrode Processes and Localized Corrosion, First Edition. Yongjun Tan.
© 2013 John Wiley & Sons, Inc. Published 2013 by John Wiley & Sons, Inc.

void formation and very poor corrosion protection properties, especially when the concentration is above the critical value [4]. Residual solvent in coating films was found to promote the formation of dark oxide spots under alkyd lacquers [5]. The existence of localized internal stress in a coating film may lead to loss of adhesion and local coating cracking [6], adding further to the complexity of inhomogeneities in coating systems. Coating inhomogeneity was also found to significantly influence the reproducibility and reliability of electrochemical evaluation of organic coatings. For example, Lee et al. used a multielectrode array to study inhomogeneities on a chemically or biologically modified electrode surface and found that they could drastically influence even the waveshape of cyclic voltammograms [7].

Knowledge of inhomogeneities in a coating or inhibitor film is thus critical for understanding underfilm corrosion processes, for evaluating corrosion prevention by coatings, and for providing insights into possible ways of improving anticorrosion coatings and inhibitors. The presence of inhomogeneities on coated electrode surfaces is usually detected by visual observation or by employing imaging techniques such as optical microscopy, scanning electron microscopy, scanning tunneling microscopy, and atomic force microscopy. Some forms of inhomogeneities, such as rusts, air bubbles, pores, and cracks, might be visually or microscopically observable; however, many other forms of inhomogeneities, such as dissolved salts, trapped solvents, internal stress, imperfect film formation, and nonbonded areas, are often invisible. To understand the processes and mechanisms of underfilm corrosion, techniques such as localized pH electrodes are used to investigate the effects of inhomogeneities on local pH changes in cathodic sites [8]. Electrical resistance measurement has also been used to study the nature of underfilm corrosion processes on mild steel immersed in a sodium chloride solution [9]. Electrochemical methods such as electrochemical impedance spectroscopy (EIS) have found widespread applications for characterizing the anticorrosion performance and the degradation mechanism of anticorrosive coatings and inhibitor films [10].

Although extensive research has been carried out over the past decades, due primarily to technological limitations in probing electrode–solution interfaces, inhomogeneity on coated surfaces is still one of the less well understood coating properties and is considered to be one of the most difficult to predict accurately [11]. Recent advances in research methodologies have enabled better understanding of inhomogeneities in surface films as a critical factor affecting underfilm corrosion. Scanning probe techniques such as the scanning Kelvin probe, scanning Kelvin probe force microscopy, the scanning reference electrode technique, the scanning vibrating electrode technique, local electrochemical impedance spectroscopy, and scanning electrochemical microscopy have been used in research aimed at understanding electrode inhomogeneity in surface films and its effects on localized electrode processes. These techniques are often used as complementary tools for understanding localized corrosion processes and mechanisms in defects and underneath coatings, in particular the mechanisms of cathodic delamination and filiform corrosion [12].

Another approach to understanding inhomogeneities in surface films and localized underfilm corrosion is the wire beam electrode (WBE) [13]. The WBE is a nonscanning probe technique that is able to visualize the processes of localized corrosion under a coating or an inhibitor film by measuring parameters from local areas of a working electrode surface, such as local resistance, corrosion potential, and galvanic current, providing spatial and temporal information on underfilm localized corrosion.

In this chapter we present an overview of laboratory techniques that are useful for examining electrode inhomogeneities in surface coatings and subsequent localized corrosion processes. Particular focus is on techniques developed based on the WBE and its complementary technologies for visualizing and characterizing electrochemical inhomogeneity and underfilm localized corrosion.

5.1 CHARACTERIZING INHOMOGENEITIES IN ORGANIC COATINGS AND INHIBITOR FILMS

Visual inspection of coated coupon surfaces can often determine the sizes and shapes of air bubbles and cracks in a coating film, while microvoids and contaminants can be observed through a more detailed examination of a coated surface using an optical and scanning electron microscope (SEM). The detection of weak areas in a coating film, such as nonbonded and imperfect film areas, often needs the assistance of accelerated testing, such as the immersion test and the salt-fog test [14]. These accelerated laboratory weathering tests intensify effects from the environment, so visible coating breakdown or corrosion sites could develop more rapidly than in naturally occurring environments. However, unfortunately, these accelerated exposure tests still often cannot, within their exposure time, visually show the negative effects of inhomogeneity on intact coated surfaces.

A variety of workers [3,8] found that most inhomogeneities in coatings are not due to pores or cracks but instead are due to inhomogeneous bonding within the polymer film. This inhomogeneity cannot be observed even using a SEM but can be detected using electric resistance measurement. They found that there is a significant difference in electric resistance between various areas of an organic coating. This was done by cutting a large coating sample into smaller pieces (e.g., 1cm $\times$ 1 cm in size) and measuring the dc resistance of each piece [3]. Some pieces of the coating sample had very low dc resistances, whereas others showed much higher resistances. They named areas of high and low resistance I (*indirect*) and D (*direct*) *films*, respectively. Normally, the film resistance for an I-type film is around 10^{10} to 10^{12} $\Omega \cdot$ cm^2 and for a D-type film is around 10^6 to 10^8 $\Omega \cdot$cm^2. They assumed that the D areas are about 75 to 250 μm in diameter and are distributed randomly across the coating surface according to Poisson's law. They also found that the metal surface under D film is very sensitive to corrosion [1,3,9]. However, Mayne's technique can only be used to determine whether a coating sample (e.g., 1 cm $\times$ 1 cm in size) has at least one D area. Nevertheless, the measurement of coating resistance after a period of immersion has been employed as a traditional method of evaluating the performance of anticorrosive coatings. Coatings that are unable to maintain a high level of electrical resistance are usually those with pinholes, low coating thickness, and other defects that allow oxygen, water, and ions to penetrate the polymer film [15–17].

Electrochemical methods have been used widely to characterize the performance of anticorrosion coatings and inhibitor films because electrochemical methods are believed to be able to detect changes in the insulating structure and electrochemical characteristics of surface films. An advantage of electrochemical methods is their ability to obtain information regarding the degradation of both coating and substrate before the degradation can be observed visually. Among the electrochemical methods most widely applied for the characterization of anticorrosion coatings is electrochemical impedance spectroscopy (EIS) [10, 18–22]. Electrode impedance is one of the most important quantities that can be measured in electrochemistry and corrosion science by purely electrical means. The usefulness of EIS in characterizing the anticorrosion performance of organic coatings lies in its ability to distinguish the individual components of a coated electrode–electrolyte interface. Analysis of EIS data using electrical equivalent circuits makes it possible to determine coating resistances, coating and electrochemical double-layer capacitors, and other parameters related to electrode–electrolyte interfacial components. The values of these interfacial parameters and their changes with coating degradation are useful for understanding the behavior and performance of coating systems. For this reason, EIS has become a very important technique which has broadened the range of underfilm corrosion phenomena that can be studied using electrochemical techniques. For example, EIS has been applied extensively to the investigation of water and ion transport in organic coatings and subsequent corrosion processes. A typical example of the use of EIS data and modeling in investigating the effects of environmental factors on an inhibitor filmed electrode–solution interface has been discussed by Tan et al. [22]. However, it should be noted that in coating studies, EIS is generally applied in a qualitative manner. Quantitative models relating EIS measurements directly to the prediction of coating and inhibitor film lifetimes has not been successful, probably due to difficulties in obtaining accurate and reproducible EIS data and in effective EIS data analysis [23].

Electrochemical noise analysis (ENA) is another electrochemical method that has been found useful in evaluating anticorrosion coatings [24,25] and inhibitor films [26]. ENA is based on measurement of the natural voltage and current fluctuations generated from coated electrodes in corrosion cells. The most useful parameter has been considered to be the noise resistance derived as the standard deviation of the voltage noise divided by the standard deviation of the current noise [26–30]. Noise resistance measurement has been found to correlate with dc resistance measurements for coated specimens and also with polarization resistance measurements for bare metal [31,32]. When applied in parallel to coated electrodes, EIS and ENA often produce similar results [33,34]. In addition to evaluating organic coatings, ENA has been proposed as a means of distinguishing various types of corrosion [35]; however, this application remains a controversial subject because there is often no consensus on a theoretical framework for interpreting electrochemical noise data.

It should be noted, however, that in principle, electrochemical methods, including EIS and ENA, are applicable only to homogeneously coated electrode surfaces. They are not able to perform direct measurement of either local dc resistance or ac impedance. This limitation can be illustrated by examining traditional methods of measuring the electrical resistance of a surface film or coating. If a coated electrode surface is heterogeneous, conventional resistance measurement using a coating electrode with a large area and a resistance meter would only measure resistance that is a mixture of contributions from many local film resistances, none of which we can evaluate independently. Obviously, there is a need for techniques that are able to measure local electrical or electrochemical parameters, such as local coating film resistance. This is an important requirement since corrosion failure in organic coating and substrate systems is often found to be initiated at a chemical or physical inhomogeneity on the coating–electrode interface. Underfilm corrosion is found to be directly beneath degradation-susceptible regions in the coating. These degradation-susceptible regions are microscopic in dimension and have properties that are different from those of the rest of the film. Although they occupy only a small fraction of the film volume, they control the corrosion-protection performance of a polymer coating.

In recent years, research has been conducted to characterize degradation-susceptible regions in coatings by mapping polymer heterogeneity using atomic force microscopy (AFM) phase imaging and nanoscale indentation [36]. AFM has been found to possess the lateral resolution needed to detect heterogeneous regions in polymer coatings that are believed to range from nano- to micrometers, and to overcome limitations associated with micro- and spectroscopic techniques, including electron microscopy, neutron and x-ray scattering, x-ray photoelectron spectroscopy, secondary-ion mass spectrometry, near-field scanning optical microscopy, and reflection optical microscopy. In a typical experiment, AFM was used to study heterogeneity in a thin film of approximately 250 nm of polystyrene and polybutadiene blends [36]. Pits were observed to reach the film–substrate interface, creating pathways that lead to corrosion of the substrate.

Another scanning probe technique that has been employed in studying the performance of anticorrosive coatings and processes occurring over coating–metal interfaces is the scanning Kelvin probe (SKP). SKP is a noninvasive, no-contact vibrating capacitor technique that is able to measure the voltage between a vibrating microelectrode and a sample with high resolution [37–42]. The SKP technique permits the mapping of Volta potential differences at buried metal–polymer interfaces. The standard SKP has played a major role in gaining a deeper understanding of cathodic delamination. For example, Stratmann and colleagues applied SKP in measuring the interfacial potential between a defect and a random location at the coating–steel interface, permitting the rate of cathodic delamination to be measured nondestructively [37–40]. It was reported that a coating-defect site and a delamination front form the anode and cathode of a galvanic cell that is distinct among SKP potential maps because of the steep increase in potential. In a typical experiment, the potential drop at the delamination front was measured to be 200 mV, the sharpness of its gradient being 30 mV/μm. A more gradual slope in the electrode potential signifies the area already delaminated [43].

SKP measurements have also been performed to investigate localized defects in organic coatings. In typical experiments, SKP measurements were performed before and after the immersion of tin-plated mild steel food-can protective coatings, a TiO_2-enriched melaminic coating, and TiO_2 and a carbon black–enriched phenolic coating in a 0.35 wt% NaCl solution at pH 4 for 120 hours [44]. Some defects were observed on the surface represented by high and localized work function variation, compared with the average value over the surface. SKP measurements have also been performed on silane-treated copper panels and a reactive sputtered TiN-coated mirror-polished steel surface. It was reported that there is a different mean work function value for the coated and bare substrates, indicating the dissimilar electrochemical activity of the various surfaces. However, standard SKP does not allow resolution high enough to probe the submicroscopic coating defects. To gain more information about the microscopic and submicroscopic processes at the delamination front, scanning Kelvin probe force microscopy (SKPFM), which combines AFM in the KPM mode, has been used to probe Volta potentials of delaminating electrode–coating interfaces with submicrometer resolution [43]. SKPFM has been shown to be an in situ technique for gaining a more detailed understanding of localized delamination processes in the microscopic and submicroscopic ranges.

A practical difficulty in the application of SKP and SKPFM to practical coated electrodes under localized corrosion is that for all SKP and SKPFM measurements, the resolution is strongly dependent on the distance between the tip and the interface between the polymer film and the metal surface. This requires the preparation of special model samples that are characterized by ultrathin polymer coatings and specially prepared defects that show a very sharp borderline in the intact coating [43]. On the other hand, the SKP and SKPFM scanning tips are only pseudoreferences since their Volta potential may vary from tip to tip, due to slight differences in the oxide covering them, or contaminants deposited on the tip during scanning. Furthermore, recent studies suggest that the nature of the polymer film has a pronounced effect on the resulting image [43]. Another limit of the SKP is considered to be the difficulty in interpretation of the experimental data [44].

Other scanning probe techniques, such as the scanning vibrating electrode technique (SVET) and local electrochemical impedance spectroscopy (LEIS), have also been used to detect inhomogeneities in coating films and associated localized corrosion damage. In a typical experiment, SVET was used to scan localized electrochemical events 80 μm over a 4.5 mm × 4 mm coated surface area immersed in a 0.005 M NaCl solution [44]. SVET was able to detect defects in a pigment-free coating. Growth of the local anode area on the paint film was calculated statistically and determined by the difference in potential gradient values between the local anode, corresponding to the defective area of painted film, and the local cathode part, corresponding to a nondefective area [44]. LEIS was also used for the detection and mapping of defects and local corrosion events in organic coatings [45]. Various types of intentional local heterogeneities, including chemical defects within the coating, such as absorbed oil, and physical defects, such as subsurface bubbles, underfilm salt deposits, pinholes, and underfilm corrosion, were detected successfully with a five-electrode LEIS system that utilizes a split microreference electrode [45].

5.2 CHARACTERIZING INHOMOGENEITY IN ORGANIC COATINGS USING THE WBE METHOD

The ability to detect the location of inhomogeneities in coating films in a quantitative manner would help identify the source of coating failure and provide insight into the mechanisms of coating degradation. A multielectrode array, the wire beam electrode (WBE), has been proposed as a means of detecting and quantifying inhomogeneities in organic coating films by measuring electrical resistance distribution over a coated electrode surface [46–49]. The WBE method involves subdividing an area of a coated surface (e.g., 1 cm^2) into many small sections and measuring the electrochemical properties of each part by means of individual sensors. Using the simple experimental setup shown in Figure 5.1, nonuniform distribution of electrical resistances over a coated WBE surface was mapped. A typical example of nonuniform distribution of coating electrical resistance is shown in Figure 5.2.

Wu et al. [49] investigated electrochemical inhomogeneities in organic coatings, in particular the D and I areas, after improving experimental condition control and high-resistance measurement techniques. An improved experimental design (Figure 5.3) was used to verify the existence of D and I areas in coating films—a significant coating characteristic first reported by Mayne et al. [1,3]. A WBE with 121 iron wires 1.0 mm in diameter was used for accurate measurement of the

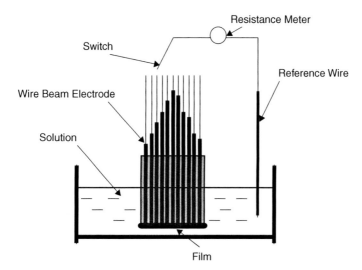

Figure 5.1 Measuring the distribution of electrical resistances in a coating film using a WBE. (From [46].)

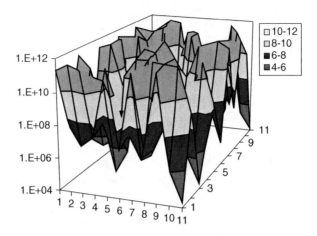

Figure 5.2 Distribution of dc resistance over a coated WBE surface. (From [49].)

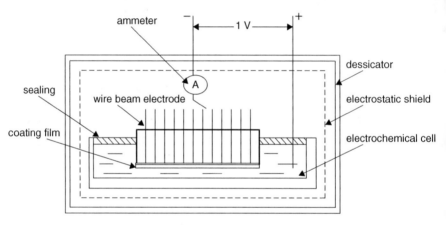

Figure 5.3 Improved experimental apparatus for mapping coating resistance distribution. (From [49].)

electrical resistance of the coating over a wide range of resistance range (10^2 to 10^{14} $\Omega \cdot cm^2$). Careful moisture control, electrostatic shielding, cable insulation, electrode surface preparation, and equipment calibration were used for accurate electrical resistance measurements. Electrical resistance measurements were carried out by applying a voltage between a cathode (a wire in the coated WBE) and an anode (an iron wire made from the same material as those in the WBE). Each terminal of the wires in the WBE was connected to the ammeter manually, in sequence, to measure the currents induced by the applied voltage. Measurements were repeated after various immersion periods in the electrolyte solution. The current measured by the ammeter was used to calculate coating film resistance using Ohm's law. In a series of experiments the inhomogeneities in three organic coatings—phenolic resin, alkyd resin, and, polyurethane varnish—were quantified by measuring the distributions of dc resistances over various surface areas of coated WBEs exposed to a 3% NaCl brine [49]. Significant differences in dc currents were recorded from different areas of coated WBE surfaces. The current values measured often show major differences between neighboring wires only 2 mm apart. For example, the maximum current measured from a typical WBE wire was 6×10^{-7} A, while the minimum current measured from another wire of the same WBE was 3×10^{-13} A. This indicates the existence of a more than 1 million-fold difference in electrical resistance over different coating areas.

Two typical types of areas identified showed a significant difference in their dc resistance. Figure 5.4 shows a typical pattern of the inhomogeneous dc resistance distributions. The two peaks' discontinuous binomial distribution, rather than a normal distribution, suggests the presence of two types of coating areas, one of higher resistance and another of lower resistance, with an obvious boundary between them. It could be evidence supporting the presence of I and D areas suggested by Mayne et al. [1,3]. Table 5.1 summarizes the resistance ranges and percentages of low and high resistance of three organic coating films. The lower-resistance areas should be covered with D films; the higher-resistance areas should be covered with I films.

The thickness of coating film was found to affect significantly the inhomogeneity of coating films. Table 5.2 summarizes coating resistance data from a coating with varying thicknesses. It can be seen that the increase in coating thickness leads to a major increase in the percentages of high-resistance coating area (I zones, 2% $\rightarrow$ 25% $\rightarrow$ 58%).

The method of coating application was also found to influence coating inhomogeneity. Table 5.3 shows results from a single- and a double-brush-painted phenolic resin coating films of 16 μm thickness. The resistance distribution of these coatings was obviously different. The percentage of high-resistance coating area for double-layer coating (38%) was larger than that of single-layer coating (25%). The boundary resistance for the double-layer coating film (10^{11} Ω) was larger than that of single-layer coating film (10^{10} Ω). This suggests that more layers can improve the corrosion-protective ability of organic coating of a certain thickness. Indeed,

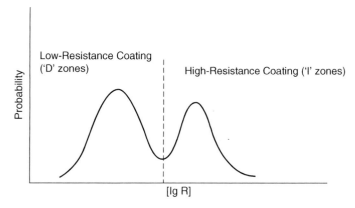

Figure 5.4   Typical dc resistance distribution plot of a coated WBE.

Table 5.1   Estimation on the Percentage of High- and Low-Resistance Films

	Phenolic Resin, 16 μm	Dry Film Thickness Alkyd Resin, 13 μm	Polyurethane Varnish, 13 μm)
Resistance range of high-resistance film (I) (Ω)	10^{10}–10^{12}	10^{10}–10^{11}	10^{10}–10^{11}
Resistance range of low-resistance film (D) (Ω)	10^{6}–10^{9}	10^{4}–10^{8}	10^{4}–10^{9}
Percent of low-resistance film (D)	75	60	58

Source: [49].

Table 5.2   Comparison of Coatings with Different Thicknesses

Thickness of Coating Film	8 μm	16 μm	25 μm
I/D coating area boundary resistance (Ω)	$\sim 10^{8}$	$\sim 10^{10}$	$\sim 10^{11}$
Percent of high-resistance film (I)	2	25	56
Percent of low-resistance film (D)	98	75	44

Source: [49].

Table 5.3   Comparison of Double- and Single-Layer Phenolic Resin-Coated WBE

Thickness of Coatings Film	Double Layers, 16 μm Total	Single Layer, 16 μm
I/D boundary resistance (Ω)	$\sim 10^{11}$	$\sim 10^{10}$
Percent of high-resistance film (I)	38	25
Percent of low-resistance film (D)	62	75

Source: [49].

some rust points were observed visually on the single-layer-coated WBE surface after 3 days' immersion in 3% NaCl brine, whereas no obvious rust was observed on the double-layer-coated WBE surface. This is in line with industrial practice: that most corrosion control coating systems need at least two coats, sometimes three or more coats. It is well known that a multiple-layer anticorrosion coating provides better protection than a single-layer coating. This could be because multiple coating layers could help cover up major imperfections in the first layer [49].

The multielectrode concept was first used in the evaluation of rustproofing oils [50] and was later used in the evaluation of the effectiveness of crevice corrosion inhibitors [51]. Wu and others employed a WBE in a series of experiments to determine the electrochemical inhomogeneity on oil-painted metal [52,53]. Their results showed that the distributions of corrosion potential and dc resistance of oil film were inhomogeneous on oil-painted metal. With the extension of exposure to corrosive media, the corrosion potential on substrate would shift in a positive direction; an area of low dc resistance could be eliminated by adding oil-soluble inhibitors [52]. They found that the repeatability and reliability of electrochemical measurements could be improved greatly by using a WBE. The protective property of organic coatings can be evaluated rapidly and based quantificationally on the distribution and probability of weak areas in a coating film [53].

Using similar experimental techniques, Zhong et al. investigated electrochemical inhomogeneity in temporarily protective oil coatings by sensing the potential variation over a WBE surface coated with preventive oil films [54–56]. It was found that the distribution of corrosion potential on the surface of oil-coated WBE was heterogeneous. When the oil film degrades, the distribution of corrosion potential was found to change from a normal probability distribution to a discontinuous binomial distribution [54]. A WBE was also used to investigate the self-repairing ability of temporarily protective oil coating. It was shown that inhibited oil coatings had the ability to self-repair, and that oil-soluble inhibitors had direct effects on the self-repairing ability of oil coating [55]. The method was also used to investigate the anticontamination performance of temporarily protective oil coatings. It was shown that salt contamination on the metal substrate had an influence on the heterogeneous distributions of corrosion potential and polarization resistance. With salt contamination, the corrosion potential distribution of oil coatings followed a discontinuous binomial probability distribution, whereas the anodic polarization resistance distribution of oil coatings was transformed from a lognormal probability distribution to an exponential probability distribution and then to a discontinuous binomial probability distribution. The cathodic polarization resistance distribution of oil coatings followed a lognormal probability distribution [56].

Typical experiments described above clearly demonstrate various applications of the WBE method in mapping inhomogeneities over coated metal surfaces by detecting coating electrical resistances and potential differences. It is possible to correlate WBE coating resistance or potential distribution maps of the type shown in Figure 5.2 with Volta potential profiles measurable using an SKP. More detailed investigations are needed in this area of research.

5.3 THE EFFECTS OF COATING INHOMOGENEITY ON ELECTROCHEMICAL MEASUREMENT

It has long been suspected that a small weakness such as a pore in a coating film could significantly affect the results of electrochemical measurement, such as EIS data [46]. Organic coating and inhibitor films are inhomogeneous in nature, and this inhomogeneity could significantly affect the reproducibility and reliability of conventional electrochemical measurement of corrosion under organic coatings.

Zhang et al. [57] carried out a study of the corrosion of steel under defective coatings in 3.5% NaCl solution using the WBE and EIS techniques. During the entire coating deterioration process, they found that the EIS diagrams were dominated by the substrate corrosion process under the defect areas, while electrochemical processes under the entire coated electrode were "averaged" out. According to the current distribution maps plotted using the WBE and EIS responses, they found that the initial high anodic and cathodic current densities were generated only at the defect areas [57]. This result suggests that nonuniform distributions of reaction and polarization currents over the electrode would affect EIS measurement.

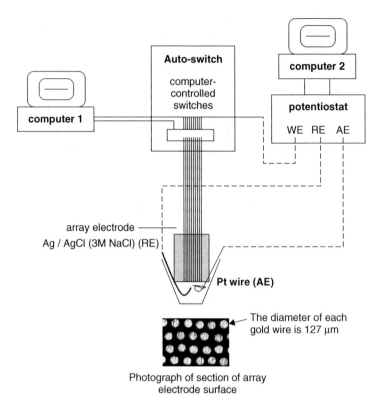

Figure 5.5 Experimental arrangement used to study the effects of electrode inhomogeneity on voltammetric responses. (From [7].) (*See insert for color representation of the figure.*)

Lee et al. carried out a sophisticated study of inhomogeneity over chemically modified electrodes and its effects on electrochemical measurement [7]. A multielectrode array consisting of 100 nominally identical and individually addressable gold disk electrodes, each with a radius of 127 μm, was used (Figure 5.5) to mimic a single macrodisk electrode in order to detect and analyze the effects of electrode inhomogeneity on voltammetric responses. The fabricated electrodes are sufficiently large that they exhibit close to linear diffusion, and each is sufficiently separated so that, with a suitable scan rate, overlap of diffusion layers can essentially be avoided. Furthermore, the individual electrodes are sufficiently small so that ohmic (*IR*) drop is minimal in studies in aqueous media. A series of experiments was performed to examine the deviation in behavior of each electrode relative to the summed response obtained when all electrodes are used simultaneously. In principle, under these circumstances, the sum of each response should equal that produced when all elements in the array electrode are operational [7]. In the investigation, the heterogeneity effect of a thiol monolayer–modified electrode surface was probed with respect to the diffusion-controlled electrochemistry of cytochrome *c*. The array configuration was initially employed with a reversible

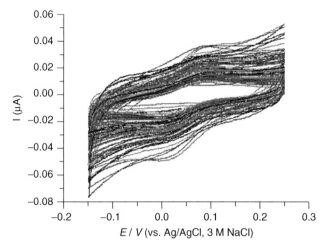

Figure 5.6 Cyclic voltammograms obtained from each 4,4′-dipyridyl disulfide–modified 127-µm-radius gold element (total of 98) of a gold multielectrode array, at a scan rate of 50 mV/s in 400µM cytochrome c (0.1 M NaCl in 20 mM phosphate buffer.) (From [7].) (*See insert for color representation of the figure.*)

and hence relatively surface-insensitive $[Ru(NH_3)_6]^{3+/2+}$ reaction and then with a more highly surface-sensitive quasi-reversible $[Fe(CN)_6]^{3-/4-}$ process. In both cases the reactants and products are solution soluble, and at a scan rate of 50 mV/s, each electrode in the array was assumed to behave independently, since no evidence of overlapping of the diffusion layers was detected.

As would be expected, the variability of the responses of the individual electrodes was significantly larger than that found for the summed electrode behavior. In the case of cytochrome c voltammetry at a 4,4′-dipyridyl disulfide–modified electrode, far greater dependence on electrode history and electrode inhomogeneity was detected. In this case, voltammograms derived from individual electrodes in the gold array electrode exhibit shape variations ranging from peak to sigmoidal (Figure 5.6). However, the total response was always found to be well defined. These results imply that random levels of inhomogeneity in gold electrode surfaces may contribute to the overall voltammetric response obtained from a gold electrode [7].

This voltammetry is consistent with a microscopic model of inhomogeneity where some parts of each chemically modified electrode surface are electroactive whereas other parts are less active. The findings are consistent with the common existence of electrode inhomogeneity in cyclic voltammetric responses at gold electrodes, which are normally difficult to detect but fundamentally important, as electrode nonuniformity can give rise to subtle forms of kinetic and other forms of dispersion. These results imply that random levels of inhomogeneities in gold electrode surfaces may contribute to the overall voltammetric response. In most cases, the influence caused by electrode heterogeneity could be subtle, although in the case of a chemically modified electrode surface, heterogeneity may drastically influence even the waveshape. This study is in agreement with studies by Compton and Banks [58] on the effects of heterogeneity at carbon electrodes and implies that electrochemists may need to recognize more widely the influence of surface inhomogeneities as a factor that introduces nonideal behavior relative to that predicted on the basis of a uniform surface [7].

A method of improving the reliability and reproducibility of coating evaluation has been proposed based on the WBE concept [47]. Electrochemical measurement and analysis using the WBE could enable the electrochemical evaluation of coated electrodes on a statistical basis, and this statistical analysis could help to avoid serious influences from random factors such as pores in a coating film on electrochemical measurement [46,47].

5.4 VISUALIZING UNDERFILM CORROSION AND THE EFFECTS OF CATHODIC PROTECTION

A WBE has been used to measure electrochemical parameters from local areas under an organic protective film, including galvanic corrosion current and corrosion potential and their distributions. These electrochemical parameters were used to study nonuniform corrosion of an electrode covered with organic coatings or films and for evaluating the corrosion protective ability of rustproof oil films. In the typical experiment [59] shown in Figure 5.7, a steel WBE was prefilmed with a rustproof oil and exposed to water drops common in corrosion under atmospheric exposure conditions. Water drops of various sizes often form on a coated metal

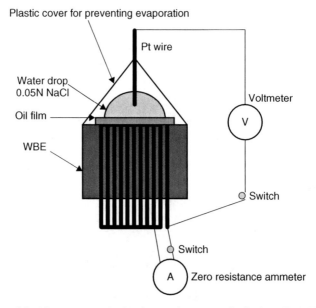

Figure 5.7 Measurements of galvanic corrosion current distribution. (From [59]).

surface and cause localized corrosion damage. This experiment used experimental designs similar to those shown in Figures 5.4 and 5.5, the only difference being that an organic film was prepainted on the working surface of the WBE before it was exposed to water-drop corrosion conditions. Two rustproof oil films were used: the first a thin film of the very widely used rustproof oil WD-40 and the second a thin film of engine oil (Mobil SAE 20W-50). The thickness of the oil films was approximately 10 μm.

Using the experimental design shown in Figure 5.7, galvanic current distributions over a WBE surface, filmed with a thin layer of rustproof oil WD-40, were measured at various stages of the exposure period. At the beginning of the exposure, as shown in Figure 5.8a, there was a small area that exhibited a large anodic

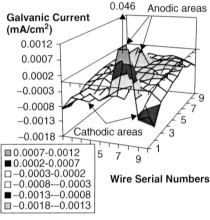

(a) After 2 hours' exposure

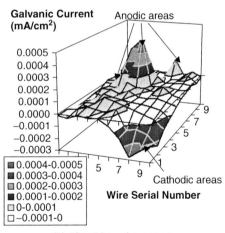

(b) After 4 hours' exposure

Figure 5.8 Galvanic current distributions over a WBE surface, with a thin layer of rustproof oil (WD-40), exposed to a drop of 0.05 N NaCl solution (approximately 12 mm in diameter). (From [59].)

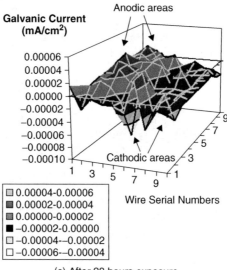

(c) After 28 hours exposure

Figure 5.8 (*Continued*)

current peak (0.046 mA/cm^2). This peak may correspond to a weak area in the oil film. However, with extended exposure this large anodic current peak disappeared and was replaced by much smaller anodic peaks (Figure 5.8b). This phenomenon may be related to the self-repair processes of this rustproof oil film, although the exact reason needs further investigation. Generally, very low galvanic currents were recorded in this system, although there was a clear separation of anodic and cathodic zones under the rustproof oil film. This result corresponds well with the good rustproof ability of this widely used rustproof oil.

Using the same experimental design, galvanic current distributions over a WBE surface, filmed with a thin layer of less protective engine oil (Mobil SAE 20 W-50), were measured at various stages of the exposure period. At the beginning of the exposure, as shown in Figure 5.9a, a large anodic current peak (0.016 mA/cm^2) was recorded from a small electrode area. This anodic current peak, however, did not disappear with extended of exposure; instead, it increased with exposure time (Figure 5.9b and c). At the end of this exposure test, brown corrosion products were observed at the anodic current peak location, which obviously corresponded to a weak area in the oil film. This oil film was not able to self-repae the weak area. This result correlated well with the less rustproof ability of this engine oil.

A WBE can also be used in conjunction with other techniques, such as scanning probes to visualize underfilm corrosion processes from both the metallic and electrolyte phases to achieve a better understanding of corrosion mechanisms. In a typical experiment, a WBE was sprayed with a layer of WD-40 oil film and exposed to Evans solution in an experimental arrangement (Figure 5.10). Figure 5.11a shows WBE and SRET maps measured immediately after the specimen was immersed in

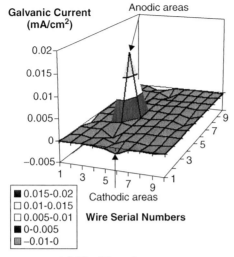

(a) After 2 hours' exposure

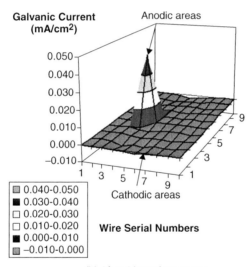

(b) After 4 hours' exposure

Figure 5.9   Galvanic corrosion current distributions over a WBE surface, with a thin layer of engine oil, exposed to a drop of 0.05 N NaCl solution (approximately 12 mm in diameter). (From [59].).

the solution. Both types of maps successfully detected anodic sites that clearly correlated with each other, although cathodic zones in SRET maps were affected by scanning tip movement [60]. At the beginning of the exposure, as shown in Figure 5.11a, there were 37 wires that behaved as anodes, although the galvanic current values detected by the WBE method were very small (the maximum anodic current was 0.068 mA/cm^2). These anodic sites are believed to be the weakness sites in the WD-40 oil layer, which can be attributed to the electrochemical

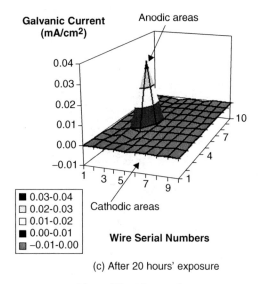

(c) After 20 hours' exposure

Figure 5.9 (*Continued*)

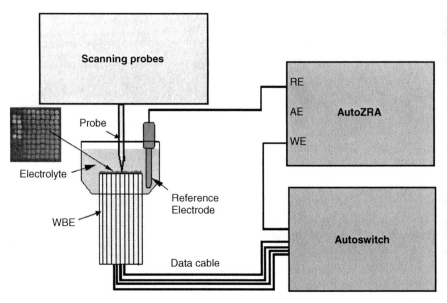

Figure 5.10 Experimental setup of a filmed WBE in combination with SRET. (From [60].)

inhomogeneity of organic coatings. An interesting observation in the experiment was that the locations of major anodic sites remained almost unchanged, but the number of anodic sites decreased considerably when the experiment was extended. After 2 hours' immersion, only 14 wires remained as anodes. The maximum galvanic current increased steadily from 0.068 mA/cm^2 to 0.481 mA/cm^2 during 20

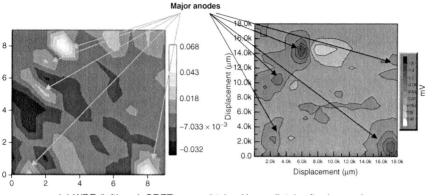

(a) WBE (left) and SRET maps obtained immediately after immersion

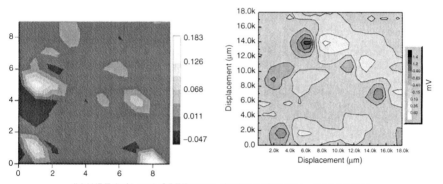

(b) WBE (left) and SRET maps obtained after 2 hours' immersion

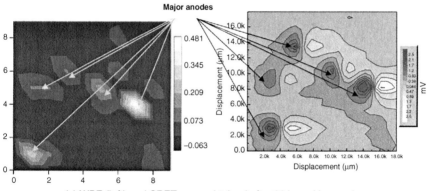

(c) WBE (left) and SRET maps obtained after 20 hours' immersion

Figure 5.11   WBE current (in mA/cm^2) and SRET maps measured from a mild steel WBE surface coated with a WD-40 oil layer and exposed to Evans solution for various periods. (From [60].)

hours' exposure, suggesting that corrosion became more and more localized and concentrated. This result is surprising since extended exposure to a corrosion environment is expected to cause continued degradation of the oil film and thus more anodic sites. The mechanism of this phenomenon requires further detailed investigation.

During the entire experimental period, the WBE and SRET map results were correlated. The WBE maps, in particular the typical potential distribution shown in Figure 5.12, appear to give finer details on the behavior of corrosion anodes and cathodes. This experiment confirms that the combined WBE–SRET method was able to provide useful information on macrocell electrochemical corrosion processes that involve macroscale separation of anodes and cathodes. The WBE–SRET method is useful for understanding the initiation, propagation, and electrochemical behavior of localized corrosion anodes and cathodes, and their dependence on externally controllable variables such as the existence of surface coatings [60].

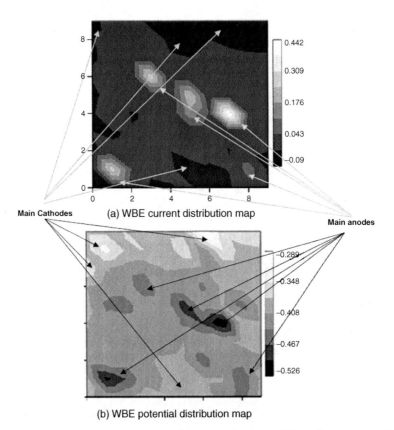

Main Cathodes   (a) WBE current distribution map

Main anodes

(b) WBE potential distribution map

Figure 5.12 WBE galvanic current (in mA/cm^2) and potential (V vs. SCE) distribution maps measured from a mild steel WBE surface coated with a WD-40 oil layer and exposed to Evans solution for 38 hours. (From [60].)

5.5 STUDYING CORROSION PROTECTION BY COATINGS AND CATHODIC PROTECTION

The use of WBEs has been extended to an evaluation of corrosion prevention techniques. In an experiment described by Tan and Wang [61], a WBE was used as a tool for monitoring the anodic electrodeposition of polyaniline (PANI) coatings and also to understand the anticorrosion performance and mechanism of PANI coatings. Anodic polarization currents were measured from various locations over the WBE surface to produce anodic polarization current maps. Experimental results revealed that if an AA1100 WBE was not pretreated, the map would show a localized anodic current distribution, resulting in a nonuniform PANI deposit. If the AA1100 electrode was pretreated using a cathodic polarization process, the map would show a random anodic current distribution, and the PANI coating would cover the entire WBE surface. These results indicated that a WBE is a practical tool for monitoring, characterizing, optimizing, and evaluating electrodeposited surface coatings such as PANI coatings [61].

Another application is the evaluation of cathodic protection current distribution over an electrode surface with and without the presence of an organic coating. In industry, cathodic protection is often used in conjunction with organic coatings to prevent localized corrosion at weak areas in the coating film. In the case of cathodic protection of a coated metal structure using a sacrificial anode, protection current (galvanic current) is not distributed uniformly over the metal structure surface. Locations that are far away from the sacrificial anode site or under a high-resistance coating could have low protection current density and thus may not be protected effectively. This is a major problem that has to be addressed when a cathodic protection system is designed. Similar problems arise when impressed cathodic protection current is applied to a metal structure such as a long pipeline with protective current density decaying as the distance to the impressed current source increases. Thus, locations far away from the current source and sites covered by high-resistance media may not be protected effectively.

In a sample experiment [59] a WBE was tested to measure the nonuniform distributions of protective current over a metal surface covered by porous organic coating film. At the end of the water-drop exposure experiment illustrated in Figure 5.9, a zinc wire with a surface area of approximately 0.015 cm^2 was introduced into the water drop, replacing the position of wire 1 in the WBE. This zinc wire behaved as a sacrificial anode to prevent localized corrosion at weak areas of the rustproof oil film. Protection current distribution was measured using the experimental design shown in Figure 5.13. Cathodic protection was thus employed, with the zinc wire behaving as a sacrificial anode. As shown in Figure 5.14, the zinc wire became the only anode in the WBE system and produced a large protection current (0.50 mA/cm^2) to protect other mild steel wires, especially those located at the weak areas of the oil film in the water drop, from further corrosion.

These preliminary studies indicate that a WBE is a practical method of studying localized electrode processes under an organic coating or rustproofing film and

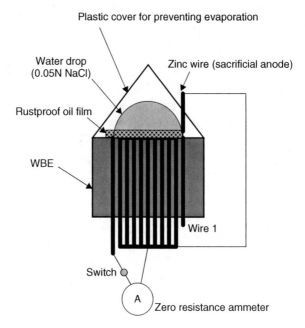

Figure 5.13 Measurements of cathodic protection current distribution. (From [59].)

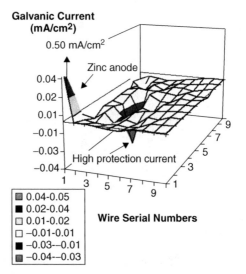

Figure 5.14 Cathodic protection current distributions over a WBE surface, filmed by a thin layer of engine oil, exposed to a drop of 0.05 N NaCl solution (approximately 12 mm in diameter.) (From [59].)

cathodic protection. A WBE can be used to determine experimentally the exact distribution of cathodic protection currents over a protected surface. This technique could provide important parameters for designing effective cathodic protection systems to avoid over- or underprotection of some sections of a coated metal structure.

Le Thu et al. [62] used a modified WBE system consisting of 210 minielectrodes to study the processes of local coating delamination in seawater under cathodic protection conditions and to evaluate compatibility between organic coatings and cathodic protection. They measured the galvanic corrosion current flowing between microelectrodes by applying a cathodic protection current. Nonuniformity of the coating was easily shown, and the delamination rate near the artificial defect was estimated. When the coating was intact, EIS revealed highly resistive behavior for 10 months, which is usually the case with thick commercial coatings devoted to cathodic protection. However, current measurements with WBE showed imperfections on the surface, which may indicate preferential delamination zones after 10 months of immersion under strong cathodic protection. The authors suggested the possibility of applying a modified WBE as an efficient method of evaluating the compatibility between organic coatings and cathodic protection.

REFERENCES

1. D. Scantlebury, The dynamic nature of underfilm corrosion, *Corrosion Science*, 35 (1993), 1363–1366.

2. N. S. Sangaj and V. C. Malshe, Permeability of polymers in protective organic coatings, *Progress in Organic Coatings*, 50 (2004), 28.

3. D. J. Mills and J. E. O. Mayne, The inhomogeneous nature of polymer films, in *Corrosion Protection by Organic Coatings*, H. Leidheiser, Jr., Ed., NACE, Houston, TX, 1981, pp. 12–17.

4. W. Funke, in *Polymeric Mute and Active Corrosion Control*, R. A. Dicke and F. L. Floyd, Eds., ACS Symposium Series 322, American Chemical Society, Washington, DC, 1986, Chap. 20.

5. W. Schwenk, in *Corrosion Control by Organic Coatings*, H. Leidheiser, Jr., Ed., NACE, Houston, TX, 1981, p. 103.

6. C. Hare, Internal stress-related coating system failures, *Journal of Protective Coatings and Linings*, 10 (1996), 99.

7. Y. C. Lee, Y. J. Tan, and A. Bond, Identification of surface heterogeneity effects in cyclic voltammograms derived from analysis of an individually addressable gold array electrode, *Analytical Chemistry*, 80 (2008), 3873–3881.

8. K. Mcleod and J. M. Sykes, in *Coatings and Surface Treatments for Corrosion and Wear Resistance*, K. N. Strsfford, P. K. Datfa, and C. G. Googan, Eds., Ellis Horwood, Chichester, UK, 1984, p. 305.

9. J. E. O. Mayne and D. J. Mills, Structural changes in polymer film on the electrolytic resistance and water uptake, *Journal of the Oil and Colour Chemists' Association*, 65 (1982), 138–142.

10. F. Mansfeld and C. H. Tsai, Determination of coating deterioration with EIS: I. Basic relationships, *Corrosion*, 47 (1991), 958.

11. G. P. Bierwagen, Reflections on corrosion control by organic coatings, *Progress in Organic Coatings*, 28 (1996), 43–48.

12. G. Grundmeier, W. Schmidt, and M. Stratmann, Corrosion protection by organic coatings: electrochemical mechanism and novel methods of investigation, *Electrochimica Acta*, 45 (2000), 2515–2533.

13. Y. J. Tan, T. Wang, T. Liu, and N. N. Aung, Studying and evaluating anti-corrosion coatings and inhibitors using the wire beam electrode method in conjunction with electrochemical noise analysis, *Anti-Corrosion Method and Materials*, 53 (2006), 30–42.

14. ASTM B117-11, *Standard Practice for Operating Salt Spray (Fog) Apparatus*, ASTM, West Conshohochen, PA, 2011.

15. E. M. Kinsella, and J. E. O Mayne, Ionic conduction in polymer films: I. Influence of electrolyte on resistance, *British Polymer Journal*, 1 (1969), 173.

16. J. E. O. Mayne and D. J. Mills, The effect of the substrate on the electrical resistance of polymer films, *Journal of the Oil and Colour Chemists' Association*, 58 (1975), 155.

17. J. E. O Mayne and J. D. Scantlebury, Ionic conduction in polymer films: II. Inhomogeneous structure of varnish films, *British Polymer Journal*, 1 (1970), 240.

18. F. Mansfeld, *Electrochemical Methods in Corrosion Testing*, ASM Handbook, Vol. 13A, ASM, Materials Park, OH, 2003, P. 445.

19. F. Mansfeld and W. J. Lorenz, Electrochemical impedance spectroscopy: application in corrosion science and technology, in *Techniques for Characterisation of Electrodes and Electrochemical Processes*, R. Varma and J. R. Selman, Eds., Wiley, New York, 1991.

20. E. P. M. van Westing, G. M. Ferrari, and J. H. W. Dewitt, The determination of coating performance with impedance measurements, *Corrosion Science*, 34 (1993), 1511.

21. G. Bierwagen, D. Tallman, J. Li, L. He, and C. Jeffcoate, EIS studies of coated metal in accelerated exposure, *Progress in Organic Coatings*, 46 (2003), 148.

22. Y. J. Tan, S. Bailey, and B. Kinsella, Investigations on the formation and destruction processes of corrosion inhibitor films using electrochemical impedance spectroscopy, *Corrosion Science*, 38 (1996), 1545–1561.

23. B. R. Hinderliter, S. G. Croll, D. E. Tallman, Q. Su, and G. P. Bierwagen, EIS studies of coated metal in accelerated exposure, *Electrochimica Acta*, 51 (2006), 4505.

24. B. S. Skerry and D. A. Eden, Electrochemical testing to assess corrosion protective coatings, *Progress in Organic Coatings*, 15 (1987), 269.

25. C. T. Chen and B. S. Skerry, Assessing the corrosion resistance of painted steel by ac impedance and electrochemical noise techniques, *Corrosion*, 47 (1991), 598.

26. Y. J. Tan, S. Bailey, and B. Kinsella, The monitoring of the formation and destruction of corrosion inhibitor films using electrochemical noise analysis, *Corrosion Science*, 38 (1996), 1681.

27. D. A. Eden, K. Hladky, D. G. John, and J. L. Dawson, Electrochemical noise resistance, Paper 274, presented at Corrosion' 86, NACE, Houston, TX, 1986.

28. F. Mansfeld and H. Xiao, Electrochemical noise analysis of iron exposed to NaCl solutions of different corrosivity, *Journal of the Electrochemical Society*, 140 (1993), 2205.

29. U. Bertocci, C. Gobrielli, F. Huet, and M. Keddam, Noise resistance applied to corrosion measurements: I. Theoretical analysis, *Journal of the Electrochemical Society*, 144 (1997), 31–37.

30. R. A. Cottis, The interpretation of electrochemical noise data, *Corrosion*, 27 (2001), 265.

31. D. Mills and S. Mabbutt, Investigation of defects in organic anti-corrosive coatings using electrochemical noise measurement, *Progress in Organic Coatings*, 39 (2000), 41.

32. D. Mills, S. Mabbutt, and G. Bierwagen, Investigation into mechanism of protection of pigmented alkyd coatings using electrochemical and other methods, *Progress in Organic Coatings*, 46 (2003), 163.

33. H. Xiao and F. Mansfeld, Evaluation of coating degradation with electrochemical impedance spectroscopy and electrochemical noise analysis, *Journal of the Electrochemical Society*, 141 (1994), 2332.

34. F. Mansfeld, L. T. Han, C. C. Lee, C. Chen, G. Zhang, and H Xiao, Analysis of electrochemical impedance and noise data for polymer coated metals, *Corrosion Science*, 39 (1997), 255.

35. C. Jeyaprabha, S. Muralidharan, G. Venkatachari, and M. Raghavan, Applications of electrochemical noise measurements in corrosion studies: a review, *Corrosion Review*, 19 (2001), 301.

36. D. Raghavan, X. Gu, T. Nguyen, M. VanLandingham, and A. Karim, Mapping polymer heterogeneity using atomic force microscopy phase imaging and nanoscale indentation, *Macromolecules*, 33 (2000), 2573–2583.

37. A. Leng, H. Streckel, and M. Stratmann, The delamination of polymeric coatings from steel: Part I. Calibration of the Kelvin probe and basic delamination mechanism, *Corrosion Science*, 41 (1999), 547.

38. A. Leng, H. Streckel, and M. Stratmann, The delamination of polymeric coatings from steel: Part 2. First stage of delamination—effect of type and concentration of cations on delamination, chemical analysis of the interface, *Corrosion Science*, 41 (1999), 579.

39. A. Leng, H. Streckel, and M. Stratmann, The delamination of polymeric coatings from steel: Part 3. Effect of the oxygen partial pressure on the delamination reaction and current distribution at the metal/polymer interface, *Corrosion Science*, 41 (1999), 599.

40. W. Fubeth and M. Stratmann, The delamination of polymeric coatings from electrogalvanized steel—a mechanistic approach: Part 2. Delamination from a defect down to steel, *Corrosion Science*, 43 (2001), 229.

41. B. Reddy and J. M. Sykes, Degradation of organic coatings in a corrosive environment: a study by scanning Kelvin probe and scanning acoustic microscope, *Progress in Organic Coatings*, 52 (2005), 280.

42. B. Reddy, M. J. Doherty, and J. M. Sykes, Breakdown of organic coatings in corrosive environment by scanning Kelvin probe acoustic microscopy, *Electrochimica Acta*, 49 (2004), 2965.

43. M. Rohwerder, E. Hornung, and M. Stratmann, Microscopic aspects of electrochemical delamination: an SKPFM study, *Electrochimica Acta*, 48 (2003), 1235–1243.

44. S. Rossi, M. Fedel, F. Deflorian, and M. C. Vadillo, Localized electrochemical techniques: theory and practical examples in corrosion studies, *Comptes Rendus Chimie*, 11 (2008), 984–994.

45. M. W. Wittmann, R. B. Leggat, and S. R. Taylor, The detection and mapping of defects in organic coatings using local electrochemical impedance methods, *Journal of the Electrochemical Society*, 146 (1999), 4071–4075.

46. Y. J. Tan, The effect of inhomogeneity in organic coatings on electrochemical measurements using a wire beam electrode: Part 1. *Progress in Organic Coatings*, 19 (1991), 89–94.

47. Y. J. Tan and S. T. Yu, The effect of inhomogeneity in organic coatings on electrochemical measurements using a wire beam electrode: Part 2, *Progress in Organic Coatings*, 19 (1991), 257–263.

48. Y. J. Tan and S. T. Yu, Study and evaluation of organic coatings by use of a wire beam electrode, *Proceedings of the 7th Asian and Pacific Corrosion Control Conference*, Beijing, 1991, p. 671.

49. C. L. Wu, X. J. Zhou, and Y. J. Tan, A study on the electrochemical inhomogeneity of organic coatings, *Progress in Organic Coatings*, 25 (1995), 379–389.

50. Y. J. Tan and X. Y., Xu, Evaluating corrosion protective oils using a multi-working electrode system, *Materials Protection* (China), 20 (1987), 38–40.

51. Y. J. Tan, A new method for crevice corrosion studies and its use in the investigation of oil-stain, *Corrosion*, 50 (1994), 266–269.

52. C. L. Wu, Q. D. Zhong, and J. C. Jin, Study on electrochemical inhomogeneity on oil painted metal, *Corrosion Science and Protection Technology* (China), No. 3, 1996.

53. G. F. Huang, G. X. Li, W. Q. Huang, C. L. Wu, and R. J. Wang, Progress in the study on organic coatings on metals by using the wire beam electrode, *Corrosion and Protection*, No. 8, 2004.

54. Q. D. Zhong, Potential variation of a temporarily protective oil coating before its degradation, *Corrosion Science*, 43 (2001), 317–324.

55. Q. D. Zhong, A novel electrochemical testing method and its use in the investigation of the self-repairing ability of temporarily protective oil coating, *Corrosion Science*, 44 (2002), 1247–1256.

56. Q. D. Zhong and Z. Zhang, Study of anti-contamination performance of temporarily protective oil coatings using wire beam electrode, *Corrosion Science*, 44 (2002), 2777–2787.

57. W. Zhang, J. Wang, Y. N. Li, and W. Wang, Evaluation of metal corrosion under defective coatings by WBE and EIS technique, *Acta Physico-Chimica Sinica*, 26 (2010), 2941–2950.

58. R. G. Compton and C. E. Banks, *Understanding Voltammetry*, World Scientific, London, 2007.

59. Y. J. Tan, Wire beam electrode: a new tool for studying localized corrosion and other heterogeneous electrochemical processes, *Corrosion Science*, 41 (1999), 229.

60. Y. J. Tan, T. Liu, and N. N. Aung, Novel corrosion experiments using the wire beam electrode: III. Measuring electrochemical corrosion parameters from both the metallic and electrolytic phases, *Corrosion Science*, 48 (2006), 53–66.

61. Y. J. Tan and T. Wang, Understanding electrodeposition of polyaniline coatings for corrosion prevention applications using the wire beam electrode method, *Corrosion Science*, 48 (2006), 2274–2290.

62. Q. Le Thu, G. Bonnet, C. Compere, H. Le Trong, and S. Touzain, Modified wire beam electrode: a useful tool to evaluate compatibility between organic coatings and cathodic protection, *Progress in Organic Coatings*, 52 (2005), 118–125.

6

Designing Experiments for Studying Localized Corrosion and Its Inhibition in Inhomogeneous Media

Over the past several decades, many corrosion testing methods, including standardized tests [1–7], have been developed and applied in various laboratories for measuring general and localized corrosion and for evaluating corrosion inhibitors. A wide range of methods are available for selection; however, in reality, corrosion research frequently needs custom-designed tests that are uniquely suited to evaluating corrosion behavior and corrosion inhibitor performance under specific environmental conditions. It is a highly challenging task to design corrosion tests, especially accelerated tests, which are able to effectively simulate corrosion behavior in actual service environments and reliably evaluate the effects on corrosion processes, rates, and mechanisms of corrosion control techniques such as corrosion inhibitors.

It is well appreciated that corrosion testing needs to simulate the actual service exposure environment; however, relatively less consideration has been given to the effects of environmental parameters on corrosion patterns and mechanisms. It is not uncommon to receive misleading test results, due to inappropriate selection of

Heterogeneous Electrode Processes and Localized Corrosion, First Edition. Yongjun Tan.
© 2013 John Wiley & Sons, Inc. Published 2013 by John Wiley & Sons, Inc.

testing parameters and measuring techniques. Another challenging issue in corrosion testing is the measurement of corrosion, especially localized forms of corrosion in highly resistive and inhomogeneous media. There are practical difficulties in setting up and maintaining corrosion testing cells in highly resistive media, problems associated with IR potential drops and nonuniform polarization current distribution, and limitations associated with localized corrosion monitoring and detection.

In this chapter we provide an overview of fundamental aspects of corrosion test design and innovative experimental methods for studying localized corrosion and for evaluating corrosion inhibitors. Attention is directed principally to technological innovations that have been made over the past decades, such as the wire beam electrode (WBE) and electrochemical noise analysis (ENA) techniques. Attempts are also made to discuss fundamental limitations in corrosion measurement techniques and challenges that may lead to the reporting of inaccurate corrosion rates and patterns. However, detailed descriptions of theoretical, experimental, and data analysis issues regarding corrosion testing techniques will not be attempted, because these have already been given by many authors, among them Mansfeld [8], Kelly et al. [9], Cottis et al. [10], Marcus and Mansfeld [11], McIntyre and Mercer [12], Baboian and Dean [13], and Tan [14].

6.1 BASIC ISSUES IN LOCALIZED CORROSION AND INHIBITOR TEST DESIGN

Acceleration of a corrosion process is usually a key requirement in corrosion tests. This is usually achieved through the intensification of major corrosion-controlling factors and the enhancement of the aggressivity of the test environment. An inappropriately accelerated corrosion test could introduce major uncertainties in test results. The identification of major environmental factors that may control the thermodynamics, kinetics, and mechanism of a corrosion process usually is the first step in successful corrosion test design. The identification of corrosion-controlling factors requires good knowledge of the nature and mechanism of a corrosion process.

The most important controlling factors for aqueous corrosion include solution composition, temperature, aeration, and flow velocity. Other major contributing factors include test duration, specimen surface conditions, the volume and mass of test solution, pressure, pH, wear, abrasion, and the existence of crevice in the test specimen. The composition and concentration of a testing solution often affect the electrolyte conductivity, the effectiveness of an electrochemical corrosion cell, and thus the rate of corrosion. In an accelerated corrosion test, for example, electrolyte concentrations are often chosen to allow a sufficient degree of corrosion in a short period of time and yet allow discrimination of the effectiveness of corrosion-control measures such as inhibitors. Possible changes in testing environment with the extension of test duration should also be taken into consideration when designing the most appropriate testing conditions. For example, the volume of a solution must be sufficiently large to avoid the exhaustion of corrosive constituents and the accumulation of corrosion products. A minimum of 250 mL of testing solution for each 6.3 cm^2 of specimen area is recommended in NACE TM0169 [6].

Temperature and aeration are critical corrosion-controlling factors. Temperature is often decided by the actual corrosion exposure environment of interest. Aeration can affect corrosion rates and patterns in different ways depending on material–environment combinations. In the case of active mild steel corrosion in neutral brine solutions, the corrosion rate–determining factor is often the diffusion of oxygen to the metal surface. Corrosion testing in a deep stagnant brine solution where the transportation of oxygen is slow would provide significantly lower corrosion rates than would testing in a shallow stirred solution. In a stirred solution where the diffusion of oxygen is no longer the rate-controlling factor, the addition of a corrosion inhibitor could become the rate-determining factor. Therefore, adding an inhibitor to a stirred solution may show a much greater effect on corrosion rates than that of adding an inhibitor to a stagnant solution. For this reason, gas bubbling and solution stirring are often used as a means of enhancing aeration for accelerating an inhibitor test and for improving inhibitor test sensitivity. In practice, however, aeration and solution stirring are often determined by the actual corrosion exposure conditions and the possibility of localized forms of corrosion under stagnant conditions. The use of rotating disk electrodes can enhance aeration and mass transportation in testing solutions. More important, the rotating disk electrode is an effective means of controlling the velocity of fluid flow over an electrode surface, creating surface shear stress. The electrochemical cell in Figure 6.1 shows the use of a rotating disk electrode to study the formation of inhibitor films in a concentrated inhibitor solution and the destruction of inhibitor films under the effects of fluid shear stress [15]. Electrochemical measurements were carried out regularly after the inhibitor-filmed electrode was transferred into an inhibitor-free brine solution and rotated at 1000 rpm.

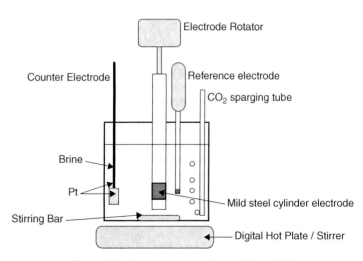

Figure 6.1   Electrochemical cell setup. (From [15].)

Corrosion mechanisms can be affected significantly by testing conditions and environmental parameters such as specimen surface conditions, wear, abrasion, time of exposure, and others. Crevices between a corrosion-testing specimen and its holders should be avoided because crevice areas are usually corrosion-sensitive sites that may not exist in actual service exposure conditions. It is important to note that different metals could respond differently to changes in environmental conditions. For example, passive metals such as stainless steel and active metals such as mild steel can respond differently to aeration and stirring of testing solutions. In the case of stainless steel exposed to neutral brine, the corrosion-controlling factor is usually the passivity of the metal surface rather than oxygen transportation. Stainless steel testing in a stagnant brine solution could record higher pitting corrosion activities than those in a stirred solution because higher oxygen concentration in a stirred solution could help to repair passive films and maintain the passivity of stainless steel surfaces. Therefore, adding corrosion inhibitors to stagnant brine solutions may result in more obvious effects on pitting corrosion than those of adding an inhibitor to a stirred solution.

A corrosion mechanism can also change with a variation in test temperature and an extension of the test duration. For example, carbon dioxide corrosion of steel at a temperature above $70°$ C can lead to the formation of a protective iron carbonate scale that can significantly affect the rate and mechanism of corrosion. If a test interval of 3 days is selected for a corrosion coupon tests in a CO_2 corrosion environment, it may lead to the reporting of falsely high corrosion rates since this short exposure test may overlook corrosion mechanism change with the formation of a protective iron carbonate scale. Extension of the coupon test to 10 days may overcome this problem; however, a 10-day test may still be far too short for microbiological buildup, which could lead to localized corrosion in actual CO_2 corrosion systems. ASTM G 31 recommends a formula for estimating suitable test duration for moderate- or low-corrosion systems: test duration (hours) = 50/(anticipated corrosion rate in mm/yr) [1]. This formula is useful; however, care needs to be taken if the corrosion mechanism could change during and after the test duration recommended.

Another important consideration when designing an accelerated corrosion test is the possibility of a change in the corrosion mechanism with adjustments under such environmental conditions as electrolyte concentration, temperature, and aeration. Although it is well appreciated that corrosion testing needs to produce the same type of corrosion (uniform, pitting, crevice, etc.) as in the service exposure; relatively less consideration has been given to the effects on corrosion mechanisms of testing conditions and environmental parameters. It is important to ensure that adjustments to the environmental conditions in accelerated tests do not change the mechanisms of corrosion. Otherwise, such tests could introduce major uncertainties in test results because the pattern and severity of corrosion are often determined by corrosion mechanisms, in particular localized forms of corrosion mechanisms. More detailed analysis of corrosion test design in various environments may be found in reports and standards scattered throughout the literature, including those by Mansfeld [8],

Kelly et al. [9], Cottis et al. [10], Marcus and Mansfeld [11], McIntyre and Mercer [12], Baboian and Dean [13], and Tan [14].

6.2 FUNDAMENTAL CONSIDERATIONS IN SELECTING CORROSION MEASUREMENT TECHNIQUES

Successful corrosion testing requires suitable measurement and analysis techniques that are able to correctly record and interpret corrosion testing data. Although many testing and monitoring techniques have been employed successfully in laboratory and field corrosion testing and research, problems have been reported in the literature. For example, Papavinasam et al. [16,17] reported significant concerns regarding some testing methodologies for evaluating corrosion inhibitors in oil and gas pipeline environments. Tan [14] analyzed fundamental limitations of some electrochemical methods, especially in the measurement of localized corrosion.

Corrosion coupons are obviously the simplest and most widely used corrosion testing tool that can be employed to determine cumulative metal thinning and localized forms of corrosion, such as pitting, crevice corrosion, weld- and heat-affected zone corrosion, and erosion corrosion. In corrosion coupon tests, metal coupons of known metallurgy, size, shape, and weight are exposed to a corrosive environment and are inspected for corrosion after a period of time (e.g., every 7 days). Corroded coupons are subjected to visual and optical or microscopic examination, weight-loss measurement, and surface analysis by various surface analytical techniques. The metallurgical phases responsible for the corrosion process are usually identified with a scanning electron microscope (SEM) coupled to an energy dispersion spectroscopy system (EDS). The surface film could be observed by an atomic force microscope (AFM), while the compounds formed on a metal surface could be analyzed by x-ray photoelectron spectroscopy (XPS). ATR–FTIR vibrational spectroscopy could be used to examine the nature of complexes attached to the metal surface without removing of the film [18]. Corrosion coupons are an excellent source of corrosion information if monitoring is carried out correctly and maintained continuously. However, corrosion coupon tests have well-known limitations; they are considered to be time consuming and may require periodic removal of the test specimen from the corrosive environment, which is cumbersome and may alter the progress of localized corrosion. They detect only the cumulative corrosion damage at the end of the exposure period and provide little information on specific events that may have triggered this damage. Although the corrosion coupon test appears to be an easy task, there are, in fact, problems that often lead to unsuccessful and misleading results. An example is the testing of underdeposit corrosion using corrosion coupons. If a corrosion coupon is fully covered by sand, it would not simulate the important galvanic corrosion effects and thus would not help to detect underdeposit corrosion problems [19,20]. Unfortunately, corrosion coupon tests are rarely questioned, as intuitively they seem to represent perfectly the corrosion occurring in an environment, and little attention has been paid to understanding why some laboratory corrosion coupon tests fail to represent field experience accurately.

An electrical resistance probe is often referred to as an "intelligent" weight-loss coupon that, in principle, should also be applicable to corrosion testing. This type of probe monitors corrosion by measuring the electrical resistance of a thin metal wire since the resistance of the wire increases as the wire becomes thinner due to corrosion dissolution. An advantage of the probe is that it provides cumulative metal loss values without the need to remove the samples from the service environment. Another advantage is that the technique is applicable to both conductive and nonconductive corrosion environments. A major disadvantage of the electrical resistance technique is that it is unable to detect localized corrosion since localized corrosion may lead to neither significant metal dissolution nor noticeable change in electric resistance. It generally does not respond rapidly to a change in corrosive conditions, and it was reported to take 4 days to respond to a 1-m/yr corrosion rate. For this reason, the electrical resistance probe is not a preferred technique for accelerated laboratory corrosion inhibitor tests. For the same reason, physics-based corrosion inspection methods, including ultrasonic and radiographical testing, are used mainly for field inspection of corrosion damage and cracks; they are rarely used in accelerated laboratory corrosion and inhibitor tests.

Localized corrosion is highly time and condition dependent; for example, some forms of localized corrosion, such as microbiologically influenced corrosion, can accelerate rapidly and grow exponentially once initiated, and it is therefore important to identify the specific time periods of corrosion rates and patterns. For this reason, "instantaneous" techniques are important for continuous measurement of the prevailing corrosion rates for corrosion assessments. Instantaneous corrosion testing and monitoring techniques are usually electrochemical in nature, including corrosion potential measurement, linear polarization resistance (LPR), electrochemical impedance spectroscopy (EIS), electrochemical noise analysis (ENA), and many others. Electrochemical techniques measure electrochemical potentials and currents that are related fundamentally to the thermodynamics and kinetics of corrosion reactions. They are often used to measure the rates of uniform corrosion, and to determine the tendency of localized corrosion, and to study a wide range of corrosion-related phenomena, such as passivation, galvanic corrosion, and sensitization effects. In typical inhibitor tests, ENA, EIS, and LPR generally produced similar polarization resistance values and similar trends during corrosion inhibitor film formation and failure processes [21]. These measurements were made based on an assumption that the electrode surface was uniform. Indeed, in principle, conventional electrochemical techniques apply only to a uniform corrosion system and have major limitations in measuring localized corrosion.

Electrochemical noise signatures were proposed to detect localized corrosion by recognizing characteristic noise patterns (often referred to as noise signatures) in the time domain [22] or in the frequency domain [23]. Although some controversial issues still exist in the interpretation of electrochemical noise data, the noise signatures are considered to be valuable indicators of localized breakdown of passive film: the incubation, initiation, propagation, and repassivation processes of localized corrosion [24]. Electrochemical noise should be useful for identifying periods when the corrosion processes become unstable, and to recognize when the probability of localized corrosion is high. However, it should be noted that the noise signatures are often difficult to identify because many electrode processes could generate similar noise patterns. On the other hand, noise analysis is unable to provide spatial information on localized corrosion.

During the past two decades, the advent of advanced physical and electrochemical techniques has facilitated substantial progress in localized corrosion and its inhibition research. The scanning Kelvin probe (SKP) and scanning Kelvin probe force microscopy (SKPFM) are probe techniques that permit mapping of topography and Volta potential distribution on electrode surfaces. SKP and SKPFM scan the electric potential just above the electrolyte over an electrode surface to detect Volta potential differences over different parts of the electrode. SKP and SKPFM have been used in experiments for studying various forms of localized corrosion processes [25–29]. Scanning electrochemical probe techniques that scan and detect local electrode potentials, galvanic currents, and local electrochemical impedances at the metal surface or metal–electrolyte interface have been developed and applied in localized corrosion research. The scanning reference electrode technique (SRET) and the scanning vibrating electrode technique (SVET) are used to probe local ionic currents flowing in the electrolyte phase by detecting small potential variations over electrode surfaces where local electrode processes occur [30,31]. Local electrochemical impedance spectroscopy (LEIS) is another scanning probe technique that has been used to map the ac impedance distribution over an electrode surface. Impedance maps obtained from LEIS measurement could detect localized electrochemical activities over a locally corroding electrode surface, while traditional ac impedance of this electrode gave little indication of its presence [32,33]. The scanning electrochemical microscope (SECM) is a tool that enables us to perform difficult tasks of detecting localized chemistry changes by means of variously designed scanning probes. The SECM is a scanning electrochemical probe that detects amperometrically surface-generated electroactive ions or molecules in the solution phase as a function of spatial location with an electrochemically sensitive or ion-selective ultramicroelectrode tip. It has been used in corrosion research [34].

Each scanning probe technique has advantages and limitations. For this reason, different techniques are often combined and applied in a synergistic manner. It should be noted that scanning probe techniques, including SRET, SVET, LEIS, and SECM, can detect ionic currents, carried by ions in the electrolyte phase, flowing over a corroding metal surface. However, they are unable to measure currents flowing at exactly the metal–solution interface. For this reason, they may not be able to detect all ionic currents accurately, especially those flowing at the metal–solution interface. Scanning probe techniques commonly operate in a relatively specific and localized area, and thus in many circumstances, the scan image does not necessarily represent the full details of an electrode process that involves different reactions occurring simultaneously over distinctively separated electrode areas. In an investigation of corrosion it is difficult to ensure that a scanning tip is correctly positioned over a pit precursor unless the precursor is generated by the probe tip itself. This implies that successful imaging of a natural pit initiation by scanning probes could depend on "luck" in experiments. Traditional optical microscopy, SEM, and EDS are often employed with scanning probe techniques to provide topographical and chemical information that is often critical in corrosion inhibitor research.

An electrochemically integrated multielectrode array, the WBE [14], has been developed as a nonscanning probe technique. As shown in Figures 3.1 to 3.5, the WBE is designed specifically to allow localized corrosion to evolve dynamically and to propagate freely on its working surface, and to detect localized corrosion and heterogeneous electrochemical parameters without disturbing electrode processes. The WBE probes have been employed in various experiments to measure electrochemical parameters such as galvanic currents flowing between anodic and cathodic sites, and local corrosion potentials from surfaces, under conditions of pitting corrosion, crevice corrosion, waterline corrosion, and nonuniform undercoating corrosion [35–48]. The WBE method has two important characteristics that make it particularly useful for studying localized corrosion and its inhibitors in complex environmental conditions: (1) a WBE can map corrosion on an instantaneous and continuous basis, providing unprecedented spatial and temporal information on localized corrosion processes occurring under deposits; and (2) a WBE is applicable to a high-resistance multiphase environment and is thus able to simulate corrosion under high-resistance deposits such as soil and concrete.

6.3   DESIGNING CORROSION TESTS IN HIGHLY RESISTIVE AND INHOMOGENEOUS MEDIA

Corrosion in highly resistive and inhomogeneous media is a major problem that frequently causes material failure in industrial and civil structures such as steel rebar corrosion in concrete buildings, light pole corrosion in soil, corrosion under thermal insulation, aircraft corrosion in the atmosphere, and underdeposit corrosion in a multiphase oil and gas flowline. Corrosion processes in inhomogeneous media often have complex mechanisms, leading to various forms of localized corrosion, such as pitting, crevice, and galvanic corrosion. Corrosion damages in highly resistive and inhomogeneous media are often hidden and concentrated in difficult-to-access areas that are buried, covered, or wrapped in concrete, deposits, or thermal insulation. To avoid major and costly teardown of industrial structures for routine inspections, electrochemical corrosion experiments have been designed and employed as management tools to acquire valuable information on hidden corrosion. However, in practice, electrochemical corrosion testing under such highly resistive and inhomogeneous conditions as multiphase fluid, atmospheric exposure, thermal insulation, soil, concrete, and deposits can be very challenging, for several important reasons.

The first issue involves the practical difficulties associated with the setup and maintenance of conventional two-or three-electrode cells for electrochemical testing of buried structures. In highly resistive and inhomogeneous media, even if electrical continuity between working, reference, and auxiliary electrodes is maintained carefully and successfully, significant IR potential drops and nonuniform polarization current distribution problems can still cause significant uncertainties in data analysis [49,50]. It has long been recognized that the nonuniform ohmic potential drop would lead to major errors in the determination of kinetic parameters [51,52] and that apparent Tafel slopes can be several times greater than the actual values, as a result of nonuniform current distribution effects [53]. This issue makes it difficult to apply electrochemical methods such as LPR and cyclic polarization to highly resistive media because of significant IR drop–induced polarization errors. The third issue involves the theoretical limitations associated with conventional electrochemical methods in simulating and measuring localized corrosion. In highly resistive and inhomogeneous media, localized corrosion is often the key process triggering corrosion failure; conventional electrochemical methods have common limitations in measuring localized electrochemical thermodynamic and kinetic parameters because they are developed based on a uniform corrosion mechanism. Relatively new methods, such as scanning probe techniques, have been developed into useful research tools for localized corrosion studies; however, scanning probes are unable to scan corrosion occurring in concrete or under solid deposits. Obviously, successful corrosion testing in highly resistive and inhomogeneous media requires technological innovations that address these issues.

One approach to overcoming practical difficulties is innovative design of miniaturized sensors or electrodes that can be embedded in, or emplaced on, various structures. Miniature sensors offer the possibility of overcoming the practical difficulties associated with the installation and withdrawal of corrosion sensors in difficult-to-access areas such as configurational and structural overlaps on an aircraft. Miniature sensors also make it convenient to set up and maintain electrochemical testing cells. For example, Glass et al. [54] designed a miniaturized corrosion sensor system incorporating individual sensing elements for obtaining environmental corrosivity information such as pH values. This innovative design includes a miniaturized electrochemical testing cell for LPR measurements believed to provide data that are useful for early warning of environmental conditions that may cause severe corrosion. LPR [54]- and EIS [55]-based miniature sensors have also been designed to detect atmospheric corrosion on coated surfaces. However, in principle, miniaturized electrochemical sensors do not solve IR potential drops and nonuniform polarization current distribution problems in highly resistive and inhomogeneous media. They still have restrictions when being applied to highly resistive solutions, although resistance compensation could be available on some systems, and require Tafel constants for corrosion rate calculations. On the other hand, these electrochemical sensors still present difficulties in the measurement of localized corrosion—the main problem in the aircraft and most other industries [14].

Another approach is the development of novel electrochemical corrosion sensors that are able to avoid IR potential drops and nonuniform polarization current problems and are able to detect localized corrosion. Anode-ladder and expansion-ring macrocell sensors [56] have been developed to monitor the corrosion risk of concrete structures. The anode-ladder system consists of six single black steel anodes, positioned 50 mm apart to prevent interactions between the anodes. The macrocell expansion-ring sensor consists of six measuring rings (anodes) separated by sealing rings as parts of the main sensor and a cathode-bar, installed in small holes drilled into the concrete structure. The basic measuring principle is to place several electrodes into a concrete structure at various depths and to measure the macrocell current, potentials, and the electrical resistance of the concrete around the sensors. This sensor system detects the depth of the critical chloride content that may initiate depassivation and corrosion, and thus the time to corrosion can be estimated, enabling the owners of buildings to initiate preventive protection measures before cracks and spalling occur. In a typical experiment, it was found that the critical chloride content reached a depth of 5 mm into a concrete specimen about 80 days after concrete placement, which caused a significant increase in the macrocell current [56]. It was also shown that high-performance concrete without the addition of chlorides recorded very low macrocell currents, lower than 15 μA, while macrocell currents in the range of 50 μA were measured when 1 to 5% chloride cement had been added to the concrete. However, it was also found that macrocell currents did not increase with chloride content, which suggested that the sensor only gave an indication of galvanic activities due to the ingress of chlorides; it did not give information on the corrosion rates of steel in a concrete structure [56].

The WBE has been used in new approaches to the design of corrosion tests in highly resistive and inhomogeneous media [57]. As discussed in previous chapters, the WBE has unique characteristics that enable the effective simulation corrosion on a continuous metal surface in highly resistive and inhomogeneous media. These characteristics are illustrated in the following typical experimental studies of localized corrosion in various highly resistive and inhomogeneous media. Figure 6.2 shows an experiment designed to evaluate CO_2 corrosion of steel in simulated oil and gas flowline environments. A WBE was used in conjunction with electrochemical noise resistance (R_n) measurements to map local electrochemical parameters, including corrosion potential, galvanic current, and electrochemical noise resistance from corroding surfaces exposed to highly resistive and inhomogeneous multiphase mixtures of hydrocarbon/and water [38]. Figure 6.3 shows typical corrosion potential, galvanic current, noise resistance, and corrosion rate distribution maps measured from a WBE exposed to a water–kerosene–CO_2 mixture for 720 hours. It is clear that over the WBE surface there is a nonuniform distribution of corrosion potential, galvanic current, and noise resistance, and thus a nonuniform corrosion rate distribution. Localized corrosion was concentrated primarily over the area where wire 10 was located, and there was a maximum potential difference of 111 mV over the electrode surface. The noise resistance values are quite large over the majority of the electrode surface (ca. 10^7 Ω) except over the areas near wire 10 (ca. 10^3 Ω). This appears to have arisen from the turbulent flow that damaged the inhibitor film in that area and thus produced a marked decrease in polarization resistance [38].

There is generally a good correlation between maps in Figure 6.3. The damaged areas clearly exhibited low noise resistance, more negative potential, and correspondingly positive galvanic current, which indicate that electrons move from this

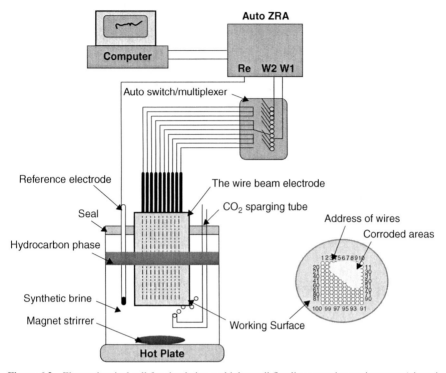

Figure 6.2 Electrochemical cell for simulating multiphase oil flowline corrosion environment (viewed with no stirring). (From [38].)

area to cathodic areas. All these data suggest that the damaged areas behaved as anodes. This is reasonable since an unprotected metallic surface would obviously form the anode of the corrosion system, and the corrosion concentrated on the anodic areas. The cathodic areas were clearly under cathodic protection. These electrochemical parameters were used to calculate the rates of localized corrosion and their distribution over a WBE surface. As shown in Figure 6.4, electrochemically calculated corrosion maps correlated clearly with microscopically observed corrosion depth maps. This experiment suggests that the combined R_n–WBE technique is capable of measuring CO_2 corrosion under multiphase conditions and that it has certain advantages over conventional electrochemical measurements for measuring corrosion in highly resistive media.

Designing tests to effectively evaluate the effects of corrosion inhibitors on localized corrosion can be highly challenging. This can be illustrated by a practical case of developing a test to evaluate the performance of CO_2 corrosion inhibitors under simulated oil flowline conditions [20]. Corrosion inhibitors are used to prevent oil pipeline failure due to pitting and mesa corrosion under solid deposits such as sand and biofilms. A problem is that the efficiency of corrosion inhibitors is often unknown because it is considered to be nearly impossible to assess by normal corrosion testing techniques [19]. Underdeposited CO_2 corrosion is believed

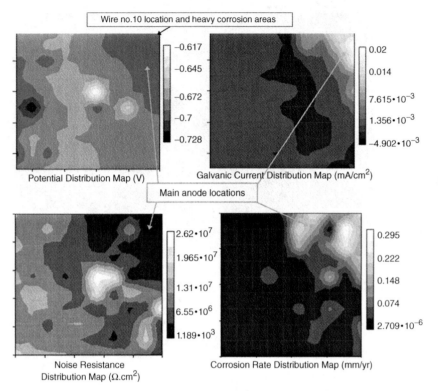

Figure 6.3 Corrosion potential, galvanic current, noise resistance, and corrosion rate maps measured from a WBE exposed to a water–kerosene–CO_2 mixture. (From [38].)

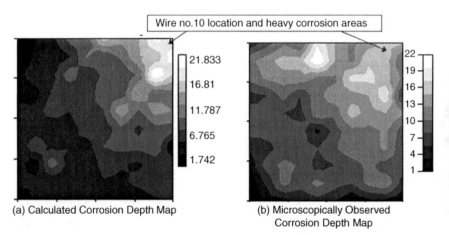

Figure 6.4 Corrosion depth maps and values (in μm) observed and calculated over a WBE surface after exposure to a water–kerosene–CO_2 mixture for 917 hours. (From [38].)

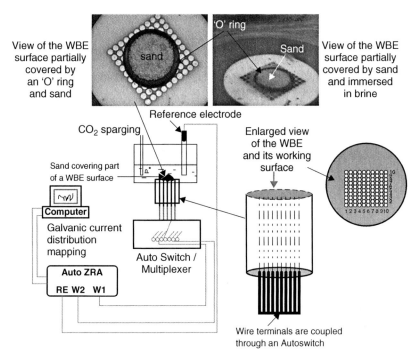

Figure 6.5   WBE test setup for underdeposit corrosion and its inhibition. (From [20].)

to be controlled by such factors as the galvanic effects between a large cathode (pipeline surface) and a small anode (surface under deposits), the failure of inhibitors to penetrate the deposits, and the retention of aggressive species in the deposits. A corrosion inhibitor test should effectively simulate these controlling factors and measure their effects on corrosion rates and patterns. Figure 6.5 shows an experimental setup where the WBE working surface is partially covered with sand and exposed to a CO_2 saturated brine solution [20]. Underdeposit corrosion occurs under the highly resistive and inhomogeneous condition that is frequently observed in oil and gas pipelines, where sand, debris, biofilm, and carbonate deposits are often present. Underdeposit corrosion is often considered nearly impossible to assess by normal corrosion testing and monitoring techniques, due to complexities in simulating underdeposit corrosion mechanisms, which, unfortunately, have not been fully addressed. Major factors thought to affect underdeposit corrosion include failure of inhibitors to penetrate the deposits, the retention of aggressive species, a large cathode/anode surface area ratio, and the formation of a localized differential "concentration cell". Development of a successful underdeposit corrosion test requires effective simulation of all these factors and related corrosion mechanisms. The WBE has been employed in a new experiment designed to simulate the underdeposit corrosion process and mechanism and to measure corrosion parameters and patterns through the detection of galvanic current and corrosion potential distributions over the WBE surface [20].

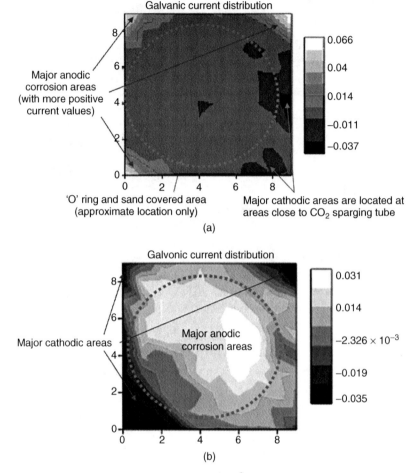

Figure 6.6   (a) Galvanic current distribution (in mA/cm^2) maps measured over a WBE exposed to a brine–CO_2 corrosion environment for 18 hours without inhibitor present; and (b) 97 minutes after 30 ppm of the inhibitor imidazoline was added to the system. (From [20].)

Figure 6.6 shows typical galvanic current distribution maps obtained from this experiment. Figure 6.6a shows galvanic current distribution over a sand-covered WBE exposed to a brine–CO_2 corrosion environment for 18 hours without inhibitor present. As shown in Figure 6.6a, corrosion anodic and cathodic sites distributed over the WBE surface with corrosion anodes were located primarily over the four corners, where no sand was present. This result suggests that underdeposit corrosion will not occur in a CO_2-saturated pure brine solution under these experimental conditions. Figure 6.6 b shows galvanic current distribution over a WBE 28 minutes after 30 ppm of inhibitor was added to the brine–CO_2 corrosion environment. As shown in Figure 6.6b, when imidazoline inhibitor was added to the solution, corrosion anodic current concentrated over the central areas of the sand deposit. This experiment illustrates the primary advantages of the WBE method:

1. It provides a means to simulate the complex localized corrosion mechanisms that often occur in highly resistive and inhomogeneous media. Electrically coupled and close-packed arrays in the WBE enable effective interactions between local anodes and cathodes through ion exchange and electron movement. This is critically important for effective simulation of underdeposit corrosion processes occurring inside and outside sand-covered areas.

2. It does not employ externally imposed polarization or a reference or counter electrode, and thus it avoids problems such as nonuniform polarization current distribution and nonuniform *IR* drops.

Figure 6.7 shows a schematic experimental setup that combines the WBE method and noise signature analysis to measure corrosion and inhibition of mild steel buried in dry, moist, and chlorinated sand [58]. To simulate different corrosion conditions, deionized water, NaCl solution, and $K_2Cr_2O_7$ solution were slowly added to the sand over various periods during the experiment. Simultaneous measurements of electrode potential noise and WBE current distribution maps were carried out. Electrode potential noise was measured by sequential measurement of the open-circuit potential of the coupled WBE system against a reference electrode. Galvanic

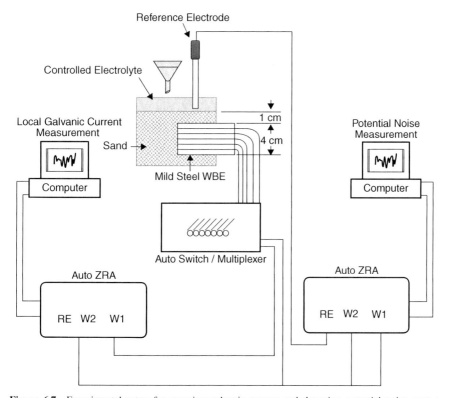

Figure 6.7 Experimental setup for mapping galvanic current and detecting potential noise over a buried WBE. (From [58].)

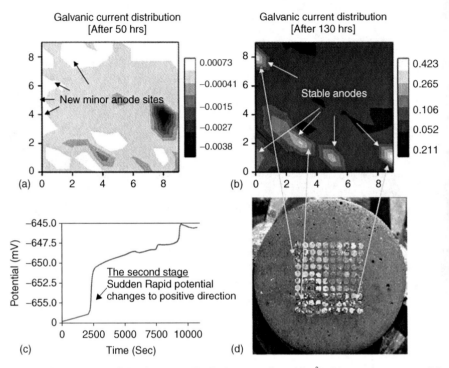

Figure 6.8 (a) and (b) Galvanic current distribution maps (in mA/cm^2); (c) second-stage potential noise; (d) WBE after completion of exposure experiment. (From [58].)

currents were mapped by connecting an AutoZRA between each wire terminal chosen and all other coupled wire terminals. The current values and their distribution in WBE current distribution maps were found to depend heavily on the type of sand being studied. Very low currents were recorded when the WBE was exposed to a dry sand environment. The current was found to increase when the sand became damp. When the WBE was buried in a chloride-contaminated sand environment, a localized corrosion pattern was observed in WBE maps. Characteristic patterns in electrode potential noise and WBE maps have been observed during the initiation and propagation of localized corrosion. Specifically, during the corrosion initiation process, many minor anodes were found to appear in the WBE current distribution maps. During the propagation of localized corrosion, anodic sites were found to disappear on a massive scale. For example, some anodes in Figure 6.8a disappeared in Figure 6.8b, which was found to correspond with the characteristic "second-stage" noise pattern of rapid potential transient to positive direction (Figure 6.8c). The maximum values of the anodic current densities increased significantly, due to the small number of remaining stable anodes, resulting in accelerated localized anodic dissolution and localized corrosion at those locations (Figure 6.8d) [58].

This experimental setup was also used in a preliminary study on the effects of corrosion inhibitors. With the addition of the corrosion inhibitor dichromate, into the sand electrode potential recording was found to shift gradually toward the negative direction and then toward the positive direction. The gradual potential shift toward the positive direction was found to correlate with contraction of the active corrosion part of the WBE. Significant reduction in maximum anodic current density values indicated that dichromate provided high corrosion-inhibition efficiency [58]. This work suggests that the WBE method can be used to study the time and spatial evolution of buried steel corrosion and the effects of corrosives and inhibitors on the corrosion process and pattern.

The main advantages of this experiment are: (1) it provides a means of simulating and monitoring complex localized corrosion processes and mechanisms occurring on buried surfaces; (2) it does not impose polarization externally and does not employ a counter electrode, and thus avoids the nonuniform polarization current distribution and nonuniform *IR* drop problems commonly associated with conventional polarization measurements; and (3) the potential noise provides useful information about dynamic events occurring on a WBE surface, which are believed to be due to localized corrosion initiation and propagation processes [58].

Corrosion under insulation is a localized corrosion problem that occurs frequently in external metal pipes when they are encapsulated in thermal insulation. Corrosion under insulation often permits degradation to proceed in an insidious manner and is thus a major concern in the petroleum, chemical, food-processing, and many other types of industrial operations. Currently, the detection of corrosion under insulation relies on methods such as radiography, ultrasonic testing, and insulation removal inspection. Traditional electrochemical techniques such as linear polarization measurements are often difficult to perform in corrosion-under-insulation studies because it is too difficult to set up a three-electrode corrosion testing cell over a metal surface that is covered by insulation of high electrical resistance. Figure 6.9 shows a schematic experiment setup in which a WBE is used in combination with noise signature analysis to monitor the penetration of corrosive species under simulated corrosion-under-insulation conditions [59]. Corrosion of an aluminum WBE exposed under insulation materials such as rock wool, glass wool, cotton wool, and tissue paper was monitored based on the observation of a typical potential noise signature. Corrosion potential noise and galvanic current distribution over the WBE surface were measured continuously to determine the time required for corrosive solution to reach the aluminum WBE and the subsequent corrosion. A typical potential noise signature observed from a AA1100 WBE is a major potential jump that corresponds to the corrosive species reaching the WBE surface, as shown in Figure 6.10. A good correlation was also observed between the WBE galvanic current maps and the corroded surface (Figure 6.11) .

Atmospheric corrosion is the most common highly resistive and inhomogeneous corrosion system and is often difficult to measure by conventional electrochemical means. A WBE has been used as a tool in experiments studying atmospheric corrosion under a small water drop [60] or a thin layer of electrolyte [61]. Figure 6.12 shows a typical experiment that was designed to study the effects of the movement of corrosive species on the process, rate, and pattern of corrosion [61]. A WBE working surface was exposed to a simulated atmospheric thin water layer where

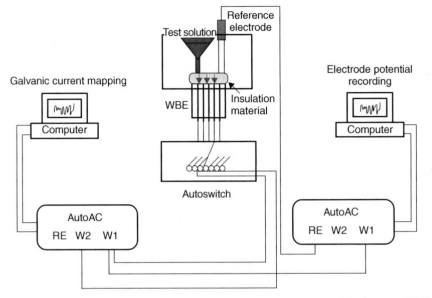

Figure 6.9 Experimental setup for mapping galvanic current and detecting potential noise over a WBE buried under insulation materials. (From [59].

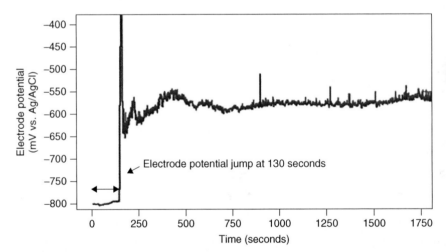

Figure 6.10 Potential noise signature obtained from a AA1100 WBE showing the potential jump, which corresponds to the corrosive species reaching the WBE surface buried in wetted rock wool. (From [59].)

there were diffusion-induced ion concentration gradients. Corrosion potential and current distribution maps were measured from WBE surfaces on a continuous basis. Typical patterns have been identified from these maps, and the characteristics of these patterns have been found to depend heavily on the type of electrode materials and the type of corrosive ions. For a mild steel WBE surface exposed to a diffusion–corrosion environment containing $NiSO_4$ or $FeCl_3$, the characteristic

Galvanic current distribution over the WBE after 3 hours

Photo of WBE corrosion after 3 hours

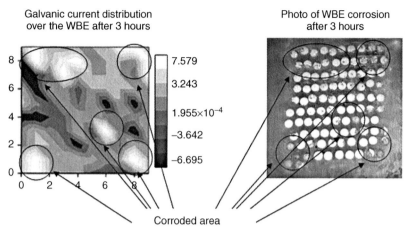

Figure 6.11 Galvanic current (in mA/cm^2) distribution map obtained from a AA1100 WBE under wetted cotton wool, and a photograph of the corroded surface. (From [59].)

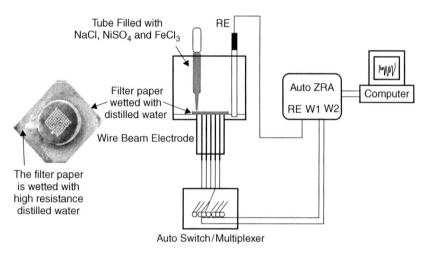

Figure 6.12 Experimental design for measuring electrochemical parameters from a WBE surface under a simulated atmospheric corrosion environment. (From [61].)

pattern in the maps shown in Figure 6.13a is found to emulate $NiSO_4$ or $FeCl_3$ concentration gradients, suggesting ion-concentration-controlled corrosion behavior. This behavior is understandable according to the Nernst equation since $NiSO_4$ diffusion would lead to both a Ni^{2+} ion concentration gradient over the WBE surface and an electrochemical potential gradient, leading to separation of anodes and cathodes over the WBE surface. When a stainless steel (SS316L) WBE surface was exposed to a diffusion–corrosion environment containing $NiSO_4$ or $NaCl$, significantly different features were observable. As shown in Figure 6.13b, corrosion cathodes and anodes appeared to be distributed randomly and, more important, these sites remained almost unchanged with the extension of diffusion–corrosion

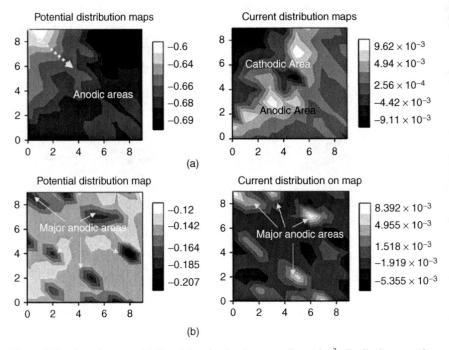

Figure 6.13 Corrosion potential (in volts) and galvanic current (in mA/cm^2) distribution maps from (a) a mild steel WBE and (b) an SS316L WBE exposed to a thin layer of distilled water after $NiSO_4$ diffusion for 1 hour. (From [61].)

experiments. The characteristic concentration-controlled corrosion patterns shown in Figure 6.13a are totally absent in Figure 6.13b. The most probable explanation of this phenomenon is that the SS316L corrosion was controlled by the nonuniformity of the passive film. It is well known that the corrosion resistance of stainless steel depends on the integrity and durability of its passive film. If passivity is not maintained and the passive film has localized weak sites or damage, these locations would become anodes where highly localized corrosion attack could occur. This work demonstrates that the recognition and analysis of characteristic maps from WBE measurements can be used as a means of studying diffusion, migration, and other forms of mass transportation–related electrochemical corrosion processes occurring under atmospheric thin-water-layer conditions.

The typical experiments described above suggest that the WBE could be utilized in conjunction with other techniques, such as ENA, to measure electrochemical corrosion in challenging and difficult-to-test corrosion conditions such as multiphase oil and gas mixtures, sand, and a thin layer of electrolyte, as well as under thermal insulation and under deposits. In all these experiments, there is no need to apply an externally imposed polarization, and there is no need of a counter electrode, and thus problems such as nonuniform polarization currents and *IR* drops associated with conventional polarization measurement are avoided.

6.4   CASE STUDIES: DESIGNING CREVICE CORROSION TESTS BY MEANS OF A WBE

Crevice corrosion is a common and insidious form of localized corrosion occurring in a fissure or occluded region, such as under washers and at welding joints. Traditionally, crevice corrosion is studied and tested in the laboratory using methods such as weight-loss measurement, visual and microscopic examination of corroded specimens, the simulated occluded cell method, and the microelectrode and freezing methods [62]. Mathematical modeling is also used extensively to study crevice corrosion. These methods can provide valuable information regarding the process and mechanism of crevice corrosion, however, they do not delineate the crevice corrosion process clearly and they do not give instantaneous crevice corrosion rates. Crevice corrosion is a heterogeneous electrochemical process in nature. When crevice corrosion occurs, there is a distinct separation of the anodic and cathodic regions on the metal surface, and different electrochemical reactions occur on the anodic and cathodic areas. To determine the instantaneous rates of crevice corrosion, electrochemical parameters at local areas of a working electrode surface, such as local anodic reaction currents, have to be determined. However, conventional electrochemical techniques present major difficulties in doing this [62].

The WBE method has been employed in various crevice corrosion tests [35,39,43]. Figure 6.14 shows an initial experimental setup for investigating

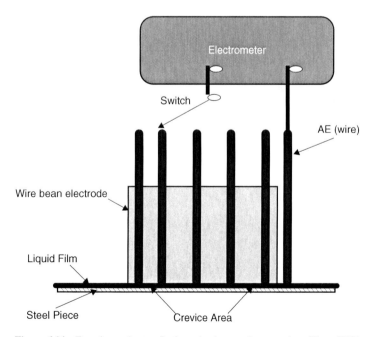

Figure 6.14   Experimental setup for investigating crevice corrosion. (From [35].)

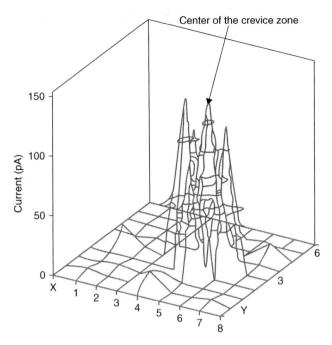

Figure 6.15   Current distribution between wires inside and outside a crevice. (From [35].)

a crevice corrosion phenomenon—oil stain—by measuring electrochemical parameters such as corrosion potential, galvanic corrosion current, and their distributions directly from crevice corrosion areas. Figure 6.15 shows galvanic current distribution in a crevice that corresponds to the occurrence of crevice corrosion and illustrates its mechanism [35].

In a series of experiments, (Figure 6.16), a WBE was used to study the corrosion of steel under CO_2 and oxygen environments in the presence of an artificial crevice [39]. Local electrochemical parameters were measured directly from crevice areas and were used to calculate local corrosion rates and their distributions. Electrochemically calculated corrosion maps were found to be quantitatively comparable to microscopic observation of a corroded electrode surface. Under an oxygen environment, localized corrosion was found to concentrate on crevice areas, whereas under a CO_2 environment, crevice corrosion was not observed.

An interesting phenomenon discovered in a CO_2 corrosion environment is that corrosion rates in a crevice area were very low. As shown in Figure 6.17, after exposure to a CO_2 corrosion environment for 1 hour, the area under the rubber band exhibited a relatively positive potential (Figure 6.17a) and very high noise resistance (Figure 6.17b). This indicates that at this stage, the electrode surface under the rubber band behaved as a cathode and corrosion rates under the rubber band were very low (Figure 6.17d), and at this initial stage, the corrosion system was not stable. At later stages of the exposure experiment, the electrode surface

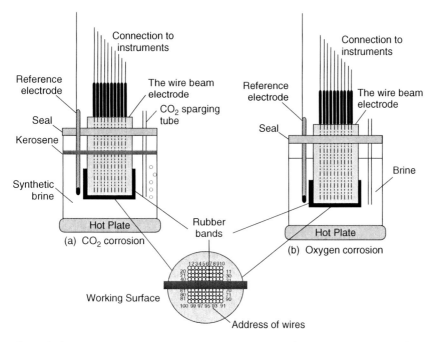

Figure 6.16 Corrosion cells for simulating various crevice corrosion environments. (From [39].)

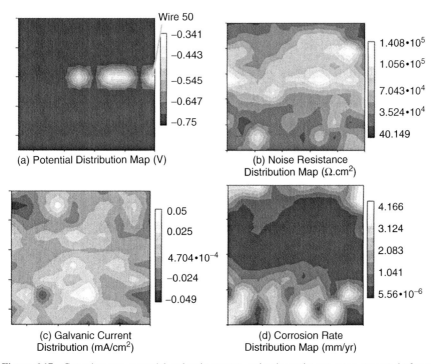

Figure 6.17 Corrosion rate, potential, galvanic current, and noise resistance maps measured after a WBE was exposed to a simulated CO_2 corrosion–crevice corrosion environment for 1 hour

under the rubber band turned into an anode and exhibited relatively more negative potential than that in other areas (Figure 6.18a). However, since noise resistance under the rubber band remained very high (Figure 6.18b), corrosion rates under a crevice remained very low (Figure 6.18d). This result suggests that for steel, crevice corrosion was not occurring under the rubber band or was occurring at very low rates. This conclusion is confirmed in Figure 6.19, showing that very little corrosion is observable in the crevice area after 502 hours' exposure. Both the microscopically observed and electrochemically calculated corrosion depth maps show very small corrosion depths ($< 1\ \mu m$) over the area where wires 41 to 50 were located, and this correlates well with the photograph of the corroded surface. This result is in agreement with the common experience that crevice corrosion is usually not an obvious problem in CO_2 corrosion environments. This crevice corrosion exemption phenomenon appears to be due primarily to a large noise resistance, which slows down corrosion in the crevice area.

Crevice corrosion behavior under an O_2 corrosion environment was completely different from that under CO_2 corrosion conditions. As shown in Figure 6.20, after exposure to an oxygen corrosion environment for only 30 minutes, clear electrochemical heterogeneity developed over the WBE surface, and the area under the rubber band behaved clearly as an anode. The corrosion potential under the rubber band was 200 to 300 mV more negative than that outside the crevice. There

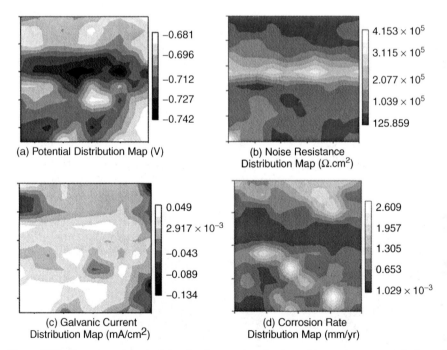

(a) Potential Distribution Map (V)

-0.681
-0.696
-0.712
-0.727
-0.742

(b) Noise Resistance
Distribution Map ($\Omega \cdot cm^2$)

4.153×10^5
3.115×10^5
2.077×10^5
1.039×10^5
125.859

(c) Galvanic Current
Distribution Map (mA/cm^2)

0.049
2.917×10^{-3}
-0.043
-0.089
-0.134

(d) Corrosion Rate
Distribution Map (mm/yr)

2.609
1.957
1.305
0.653
1.029×10^{-3}

Figure 6.18 Corrosion rate, potential, galvanic current, and noise resistance maps measured after a WBE was exposed to a simulated CO_2 corrosion–crevice corrosion environment for 138 hours

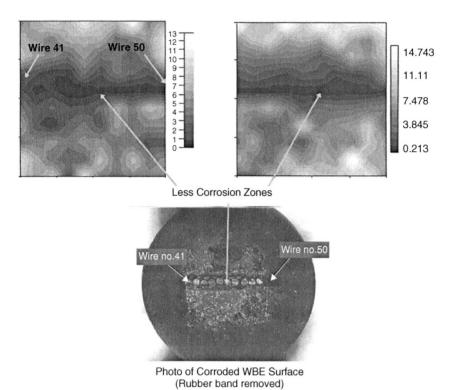

Figure 6.19   Corrosion depth maps and values (in μm) observed and calculated, as well as photographs over a WBE surface after exposure to a simulated crevice–CO_2 corrosion environment for 502 hours. (From [39].)

were large galvanic currents flowing between the inside and outside of the crevice area, resulting in rapid corrosion in the crevice. This is probably the initiation stage of crevice corrosion. At the later stages of the exposure experiment, typically as shown in Figure 6.21, crevice corrosion propagated and expanded over the rubber band area. The corrosion center moved gradually from the location of wire 31 to that of wire 40 and expanded around the crevice area while the areas outside the crevice were under cathodic protection. Over the 234-hour exposure period, the crevice area remained as an anode, and the potential difference between the anodic and cathodic areas was always above 200 mV. As a cumulative result (Figure 6.22), O_2 corrosion was concentrated in the area under the rubber band. The observed (Figure 6.22a) and calculated (Figure 6.22b) corrosion depth maps and corrosion profile photographs (Figure 6.22c) all confirm the occurrence of crevice corrosion and all show good correlation. However, a corrosion rate map calculated using the R_n–WBE technique (Figure 6.22d) does not correlate well with the corrosion depth map observed, although it also indicates that heavy corrosion occurred under the crevice area. The reason for this problem is believed to be that the noise level in the stagnant oxygen corrosion environment was too low [39].

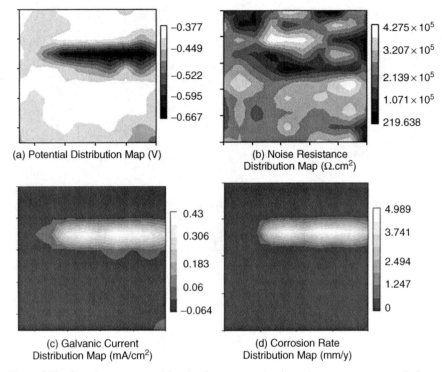

(a) Potential Distribution Map (V)

(b) Noise Resistance
Distribution Map ($\Omega.cm^2$)

(c) Galvanic Current
Distribution Map (mA/cm^2)

(d) Corrosion Rate
Distribution Map (mm/y)

Figure 6.20 Corrosion rate, potential, galvanic current, and noise resistance maps measured after a WBE was exposed to a stagnant oxygen–crevice corrosion environment for 0.5 hour.

These experiments confirm the applicability of the WBE method in crevice corrosion studies and show that the occurrence of localized corrosion depends heavily on the corrosion environment.

6.5 CASE STUDY: DESIGNING EXPERIMENTS FOR LOCALIZED CORROSION INHIBITOR DISCOVERY

A major obstacle in new inhibitor discovery is the difficulty in evaluating localized corrosion inhibition. Traditional methods of evaluating localized corrosion inhibitors are to directly measure the size and depth of each pit on a corrosion coupon microscopically. Corrosion coupon tests detect only the cumulative corrosion damage at the end of a long exposure period and do not provide dynamic pictures of localized corrosion processes. Electrochemical techniques can be used to measure general corrosion on a continuous basis, however, they are unable to measure localized corrosion rates and patterns. Scanning probe techniques enabled high-resolution observation of localized corrosion-induced events; for example, a scanning vibrating electrode has been used to monitor the kinetics of pitting corrosion as a function of inhibitor concentration [63]. However, scanning probes may

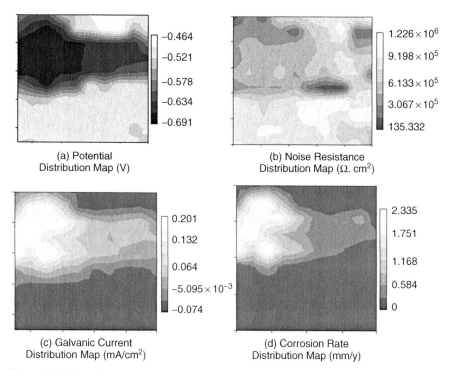

Figure 6.21 Corrosion rate, potential, galvanic current, and noise resistance maps measured after a WBE was exposed to a stagnant oxygen–crevice corrosion environment for 216 hours.

not necessarily be used effectively on surfaces that have a high level of intrinsic roughness, obscured by corrosion products, or have a large surface area. They commonly operate in a relatively small area, in many circumstances it is difficult to ensure that a scanning tip is positioned correctly over a locally corroding area.

In a typical research project [64], the WBE method was used in the search for new types of localized corrosion inhibitors. As shown in Figure 6.23, the WBE was used to study the behavior of imidazoline and an acid functionalized resorcinarene as steel corrosion inhibitors in CO_2–saturated brine solutions. Both imidazoline and resorcinarene acid were found to provide excellent inhibition to general CO_2 corrosion; however, imidazoline was found to aggravate localized corrosion by creating a small number of major anodes that focused on a small area of the WBE surface, leading to highly concentrated anodic dissolution. As shown on galvanic current distribution maps in Figure 6.24, when imidazoline was added to the corrosion cell, the number of anodic sites that display positive current values was reduced with the extention of corrosion exposure. In fact, most of the anodes disappeared and only a few major anodes formed and remained after 6 hours of exposure. These major anodes remained at the same locations on the WBE surface and did not disappear even when imidazoline concentration was increased from 30 ppm to 50 ppm.

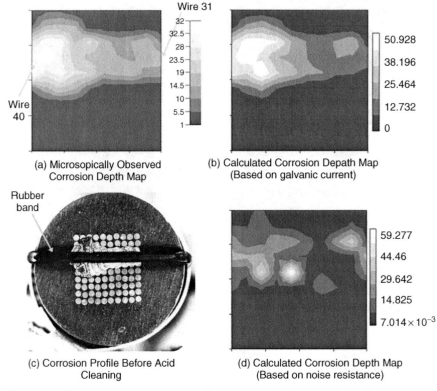

(a) Microsopically Observed Corrosion Depth Map

(b) Calculated Corrosion Depath Map (Based on galvanic current)

(c) Corrosion Profile Before Acid Cleaning

(d) Calculated Corrosion Depth Map (Based on noise resistance)

Figure 6.22 Corrosion depth maps and values (in μm) observed and calculated, as well as photographs over a WBE surface after exposure to a crevice–O_2 corrosion environment for 234 hours. (From [39].)

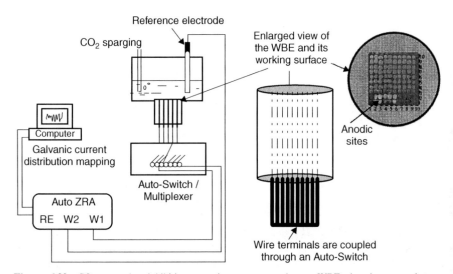

Figure 6.23 CO_2 corrosion–inhibition experiment setup using a WBE that is exposed to a CO_2-saturated brine solution. (From [64].)

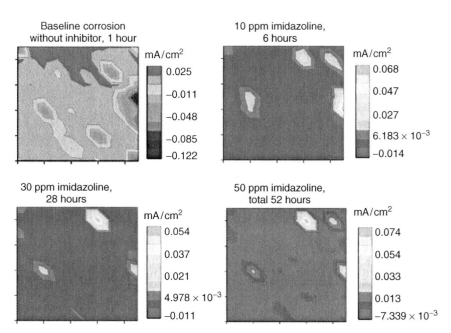

Figure 6.24 Galvanic current distribution maps measured over a WBE surface during 52 hours' exposure in brine with imidazoline present (current in mA/cm^2). (From [64].) (*See insert for color representation of the figure.*)

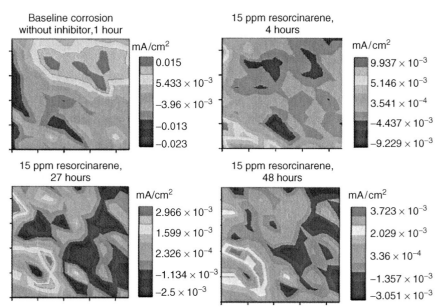

Figure 6.25 Galvanic current distribution maps measured over a WBE surface during 48 hours' exposure in brine with 15 ppm of resorcinarene acid present (current in mA/cm^2). (From [64].) (*See insert for color representation of the figure.*)

The resorcinarene acid showed distinctively different behavior by generating a large number of minor anodes distributed randomly over the WBE surface, leading to insignificant general anodic dissolution. As shown in Figure 6.25, the WBE current distribution maps are characterized by a large number of minor anodes distributed randomly over the WBE surface. Another characteristic is that the locations of most anodes kept changing in a random manner. These results indicate that resorcinarene acid provided effective localized corrosion inhibition by promoting a random distribution of insignificant anodic currents [64]. This result is significant because it suggests that organic molecules could be used as localized corrosion inhibitors if they form an interfacial structure that promotes random distribution of anodic currents.

REFERENCES

1. ASTM G31-72, *Standard Practice for Laboratory Immersion Corrosion Testing of Metals*, ASTM, West Conshohocken, PA, 2004.

2. ASTM G1-03, *Standard Practice for Preparing, Cleaning, and Evaluating Corrosion Test Specimens*, ASTM, West Conshohocken, PA, 2003.

3. ASTM G3-89, *Standard Practice for Conventions Applicable to Electrochemical Measurements in Corrosion Testing*, ASTM, West Conshohocken, PA, 2010.

4. ASTM G78-01, *Standard Guide for Crevice Corrosion Testing of Iron-Base and Nickel-Base Stainless Alloys in Seawater and Other Chloride-Containing Aqueous Environments*, ASTM, West Conshohocken, PA, 2007.

5. ASTM G44-99, *Standard Practice for Exposure of Metals and Alloys by Alternate Immersion in Neutral 3.5% Sodium Chloride Solution*, ASTM, West Conshohocken, PA, 2005.

6. NACE TM0169-2000, *Standard Test Method: Laboratory Corrosion Testing of Metals*, NACE, Houston, TX, 1995.

7. NACE Standard RP0775-2005, *Preparation, Installation, Analysis, and Interpretation of Corrosion Coupons in Oilfield Operations*, NACE, Houston, TX, 2005.

8. F. Mansfeld, *Electrochemical Methods in Corrosion Testing*, ASM Handbook, Vol. 13A, ASM, Materials Park, OH, 2003, p. 445.

9. R. G. Kelly, J. R. Scully, D. W. Shoesmith, and R. G. Buchheit, *Electrochemical Techniques in Corrosion Science and Engineering*, Marcel Dekker, New York, 2003.

10. R. A. Cottis, S. Turgoose, and R. Newman, *Corrosion Testing Made Easy: Electrochemical Impedance and Noise*, NACE International, Houston, TX, 1999.

11. P. Marcus and F. Mansfeld, Eds., *Analytical Methods in Corrosion Science and Engineering*, Taylor & Francis, London, 2006.

12. P. J. McIntyre and A. D. Mercer, Corrosion testing and determination of corrosion rates, in *Shreir's Corrosion*, Elsevier, Oxford, Vol. 2, 2010, p. 1443.

13. R. Baboian and S. W. Dean, *Corrosion Testing and Evaluation: Silver Anniversary Volume*, ASTM STP 1000-EB, ASTM, Philadelphia, 1990, p. 436.

14. Y. J. Tan, Sensing electrode inhomogeneity and electrochemical heterogeneity using an electrochemically integrated multielectrode array, *Journal of the Electrochemical Society*, 156 (2009), C195–C208.

15. Y. J. Tan, S. Bailey, and B. Kinsella, Investigations on the formation and destruction processes of corrosion inhibitor films using electrochemical impedance spectroscopy, *Corrosion Science*, 38 (1996), 1545–1561.

16. S. Papavinasam, R. W. Revie, M. Attard, A. Demoz, and K. Michaelian, Comparison of laboratory methodologies to evaluate corrosion inhibitors for oil and gas pipelines, *Corrosion*, 59 (2003), 897–912.

17. S. Papavinasam, R. W. Revie, M. Attard, A. Demoz, and K. Michaelian, Comparison of techniques for monitoring corrosion inhibitors in oil and gas pipelines, *Corrosion*, 59 (2003), 1096–1111.

18. F. Blin, S. G. Leary, G. B. Deacon, P. C. Junk, and M. Forsyth, The nature of the surface film on steel treated with cerium and lanthanum cinnamate based corrosion inhibitors, *Corrosion Science*, 48 (2006), 404–419.

19. NACE Task Group TG 380, Under Deposit Corrosion (UDC) Testing and Mitigation Methods, draft report distributed to members for review and comment, March 2009.

20. Y. J. Tan, Y. Fwu, and K. Bhardwaj, Electrochemical evaluation of underdeposit corrosion and its inhibition using the wire beam electrode method, *Corrosion Science*, 53 (2011), 1254–1261.

21. Y. J. Tan, S. Bailey, and B. Kinsella, The monitoring of the formation and destruction of corrosion inhibitor films using electrochemical noise analysis, *Corrosion Science*, 38 (1996), 1681.

22. K. Hladky and J. L. Dawson, The measurement of pitting corrosion using electrochemical noise, *Corrosion Science*, 21 (1981), 317–322.

23. K. Hladky and J. L. Dawson, The measurement of corrosion using electrochemical $1/f$ noise, *Corrosion Science*, 22 (1982), 231–237.

24. R. A. Cottis, The interpretation of electrochemical noise data, *Corrosion*, 27 (2001), 265.

25. P. Schmutz and G. S. Frankel, Characterization of AA2024-T3 by scanning Kelvin probe force microscopy, *Journal of the Electrochemical Society*, 145 (1998), 2285–2295.

26. J. H. W. de Wit, Local potential measurements with the SKPFM on aluminum alloys, *Electrochimica Acta*, 49 (2004), 2841–2850.

27. M. Jönsson, D. Thierry, and N. LeBozec, The influence of microstructure on the corrosion behavior of AZ91D studied by scanning Kelvin probe force microscopy and scanning Kelvin probe, *Corrosion Science*, 48 (2006), 1193–1208.

28. W. Fubeth and M. Stratmann, The delamination of polymeric coatings from electrogalvanized steel—a mechanistic approach: Part 2: Delamination from a defect down to steel, *Corrosion Science*, 43 (2001), 229.

29. B. Reddy and J. M. Sykes, Degradation of organic coatings in a corrosive environment: a study by scanning Kelvin probe and scanning acoustic microscope, *Progress in Organic Coatings*, 52 (2005), 280.

30. H. S. Isaacs, The use of the scanning vibrating electrode technique for detecting defects in ion vapor-deposited aluminum on steel, *Corrosion*, 43 (1987), 594–598.

31. H. N. McMurry, Localized corrosion behavior in aluminum–zinc alloy coatings investigated using the scanning reference electrode technique, *Corrosion*, 57 (2001), 313–322.

32. R. S. Lillard, P. J. Moran, and H. S. Isaacs, A novel method for generating quantitative local electrochemical impedance spectroscopy, *Journal of the Electrochemical Society*, 139 (1992), 1007–1012.

33. M. W. Wittmann, R. B. Leggat, and S. R. Taylor, The detection and mapping of defects in organic coatings using local electrochemical impedance methods, *Journal of the Electrochemical Society*, 146 (1999), 4071–4075.

34. A. Davoodi, J. Pan, C. Leygraf, and S. Norgren, Integrated AFM and SECM for in situ studies of pitting corrosion of Al alloys, *Electrochimica Acta*, 52 (2007), 7697–7705.

35. Y. J. Tan, A new crevice corrosion testing method and its use in the investigation of oil stain, *Corrosion*, 50 (1994), 266.

36. Y. J. Tan, Monitoring localized corrosion processes and estimating localized corrosion rates by means of a wire beam electrode, *Corrosion*, 54 (1998), 403–413.

37. Y. J. Tan, S. Bailey, B. Kinsella, and A. Lowe, Mapping corrosion kinetics using the wire beam electrode in conjunction with electrochemical noise resistance measurements, *Journal of the Electrochemical Society*, 147 (2000), 530–540.

38. Y. J. Tan, S. Bailey, and B. Kinsella, Mapping non-uniform corrosion using the wire beam electrode method: I. Multi-phase carbon dioxide corrosion, *Corrosion Science*, 43 (2001), 1905–1918.

39. Y. J. Tan, S. Bailey, and B. Kinsella, Mapping non-uniform corrosion using the wire beam electrode method: II. Crevice corrosion and crevice corrosion exemption, *Corrosion Science*, 43 (2001), 1919–1929.

40. Y. J. Tan, Corrosion science: a retrospective and current status, presented at The Electrochemical Society 201st Meeting, Philadelphia, G. S. Frankel, H. S. Isaacs, J. R. Scully, and J. D. Sinclair, Eds., 2002, PV2002-13.

41. Y. J. Tan, N. N. Aung, and T. Liu, Novel corrosion experiments using the wire beam electrode: I. Studying electrochemical noise signatures from localized corrosion processes, *Corrosion Science*, 48 (2006), 23–78.

42. N. D. Budiansky, F. Bocher, H. Cong, M. F. Hurley, and J. R. Scully, Use of coupled multi-electrode arrays to advance the understanding of selected corrosion phenomena, *Corrosion*, 63 (2007), 537–554.

43. F. Bocher, F. Presuel-Moreno, N. D. Budiansky, and J. R. Scully, Coupled multielectrode investigation of crevice corrosion of AISI316 stainless steel, *Electrochemical and Solid State Letters*, 10 (2007), C16–C20.

44. H. B. Cong and J. R. Scully, Use of coupled multielectrode arrays to elucidate the pH dependence of copper pitting in potable water, *Journal of the Electrochemical Society*, 157 (2010), C36–C46.

45. A. Naganuma, K. Fushimi, K. Azumi, H. Habazaki, and H. Konno, Application of the multichannel electrode method to monitoring of corrosion of steel in an artificial crevice, *Corrosion Science*, 52 (2010), 1179–1186.

46. A. Legat, Monitoring of steel corrosion in concrete by electrode arrays and electrical resistance probes, *Electrochimica Acta*, 52 (2007), 7590–7598.

47. D. L. Zhang, W. Wang, and Y. Li, An electrode array study of electrochemical inhomogeneity of zinc in zinc/steel couple during galvanic corrosion, *Corrosion Science*, 52 (2010), 1227–1284.

48. W. Wang, Y. H. Lu, Y. Zou, X. Zhang, and J. Wang, The heterogeneous electrochemical characteristics of mild steel in the presence of local glucose oxidase: a study by the wire beam electrode method, *Corrosion Science*, 52 (2010), 810–816.

49. F. Mansfield, Don't be afraid of electrochemical techniques—but use them with care! *Corrosion*, 44 (1988), 856–868..

50. J. R. Scully, Polarization resistance method for determination of instantaneous corrosion rates, Corrosion, 56 (2000), 199–218.

51. W. Tiedemann, J. Newman, and D. N. Bennion, Errors in measurements of electrode kinetics caused by nonuniform ohmic-potential drop to a disk electrode, *Journal of the Electrochemical Society*, 120 (1973), 256–258.

52. G. A. Prentice and C. W. Tobias, A survey of numerical methods and solutions for current distribution problems, *Journal of the Electrochemical Society*, 129 (1982), 72–78.

53. A. A. Sagues and S. C. Kranc, On the determination of polarization diagrams of reinforcing steel in concrete, *Corrosion*, 48 (1992), 624–633.

54. R. S. Glass, J. Clarke, L. Willis, and D. R. Ciarlo, Method for monitoring environment and corrosion, U.S. Patent 5,437,773 (1995).

55. G. D. Davis and C. M. Dacres, Electrochemical sensors for evaluating corrosion and adhesion on painted metal structures, U.S. Patent 5,859,537 (1999).

56. M. Raupach and P. Schießl, Macrocell sensor systems for monitoring of the corrosion risk of the reinforcement in concrete structures, *NDT & E International*, 34 (2001), 435–442.

57. Y. J. Tan, Experimental methods designed for measuring corrosion in highly resistive and inhomogeneous media, *Corrosion Science*, 53 (2011), 1145–1155.

58. N. N. Aung and Y. J. Tan, A new method of studying buried steel corrosion and its inhibition using the wire beam electrode, *Corrosion Science*, 46 (2004), 3057–3067.

59. N. N. Aung, W. K. Wai, and Y. J. Tan, A novel electrochemical method for monitoring corrosion under insulation, *Anti-Corrosion Methods and Materials*, 53 (2006), 175–179.

60. Y. J. Tan, Wire beam electrode: a new tool for studying localized corrosion and other heterogeneous electrochemical processes, *Corrosion Science*, 41 (1999), 229–247.

61. N. N. Aung, Y. J. Tan, and T. Liu, Novel corrosion experiments using the wire beam electrode: II. Monitoring the effects of ions transportation on electrochemical corrosion processes, *Corrosion Science*, 48 (2006), 39–52.

62. N. Corlett, L. E. Eiselstein, and N. Budiansky, Crevice corrosion, in *Shreir's Corrosion*, Elsevier, Oxford, Vol. 2, 2010, pp. 753–771.

63. G. Williams, A. J. Coleman, and H. N. McMurray, Inhibition of aluminum alloy AA2024-T3 pitting corrosion by copper complexing compounds, *Electrochimica Acta*, 55 (2010), 5947–5958.

64. Y. J. Tan, M. Mocerino, and T. Paterson, Organic molecules showing the characteristics of localized corrosion aggravation and inhibition, *Corrosion Science*, 53 (2011), 2041–2045.

7

Sensing Localized Electrodeposition and Electrodissolution

Localized electrodeposition and electrodissolution are fundamental phenomena associated with electrochemical engineering processes such as electroplating, electrowinning, electroforming, electromachining, electropolishing, and localized corrosion. Localized electrodeposition and electrodissolution are due to nonuniform distribution of electrochemical reaction currents over an electrode surface, which is considered to be undesirable in electroplating because it often results in an unevenly deposited metallic coating that cannot meet functional and dimensional requirements. When electroplating is used for depositing anticorrosion coatings, nonuniform distribution of electrodeposition current could result in burned deposits or an unevenly deposited electrode surface that has poor corrosion resistance. When electroplating is used for integrated-circuit copper metallization, uneven plating rates could result in the formation of dendrites that can sometimes stretch across the electrolyte and create a short circuit. In hydrometallurgy a nonuniform electrowinning current could lead to the formation of poorly defined metals, such as aluminum, copper, and cobalt, with undesirable mechanical and structural properties. This is particularly important for cobalt since cobalt electrowinning tends to produce deposits of high internal stress, which can cause significant problems. In aluminum extraction, a nonuniform electrowinning current can cause aluminum "metal fog" that increases the electronic conductivity of the electrolyte and reduces electrowinning cell efficiency [1,2].

Localized electrodeposition and nonuniform electrodissolution can also play desirable roles in industrial processes. For example, in electrochemical fabrication,

---

Heterogeneous Electrode Processes and Localized Corrosion, First Edition. Yongjun Tan.
© 2013 John Wiley & Sons, Inc. Published 2013 by John Wiley & Sons, Inc.

nonuniform distribution of electrodeposition current is preferred since a highly focused electrodeposition current is required for creating a fine metallic structure of accurate dimensions. Localized electrochemical deposition has been employed as a three-dimensional micro-rapid microfabrication process capable of producing extremely high-aspect-ratio microstructures [3–5]. In a typical application, nickel columns were formed electrochemically on copper cathodes from a nickel plating solution using a nonsoluble microelectrode as the anode [4]. Micrometer-scale nickel structures, including a multicoiled helical spring, were fabricated by means of localized electrodeposition by placing a sharp-tipped electrode in a plating solution, very close to a substrate, and applying a voltage of typically 4 to 5 V. Structures are built by moving the electrode appropriately with respect to the substrate [5]. The scanning electrochemical microscope (SECM) has also been used for the electrodeposition of micrometer-scale copper structures [6]. Copper columns were deposited on different substrates and interconnects within an integrated-circuit package. The potential, the concentration of the $CuSO_4$, and the presence of organic additives were all found to affect the microstructure of the deposits and control the efficiency of the deposition [6]. Localized electrodeposition has also been employed for depositing nanoparticles into higher-ordered structures using chitosan to confer spatial selectivity to electrodeposition [7,8]. Although localized electrodeposition has been used as a promising path to inexpensive free-form microfabrication in metal, techniques to create three-dimensional microstructures are still a matter of research [5]. Further investigation of localized electrodeposition is needed to achieve improved deposition control and enhanced resolution and surface smoothness.

Localized electrodissolution occurs most frequently in localized corrosion. It is also used widely in various industrial processes; for example, electromachining is based on selective dissolution of a metal workpiece by a nonuniform anodic current with a preshaped cathode "tool." Electropolishing is a procedure in which nonuniform anodic current is used selectively to dissolve microelevations on a metal surface to create a smooth surface and in some instances, a mirror finish.

In all these engineering applications of electrodeposition and dissolution, understanding of the effective control of electrochemical current distribution is a critical requirement. Description of theoretical issues of electrodeposition and dissolution have been discussed by many authors. For example, Lowenheim [1] presented a general introduction to electrodepositing. Oniciu and Muresan [9] authored a review on fundamental aspects of leveling and brightening in the electrodeposition of metals, including the effects and synergism of additives, and polarization of the cathode. Datta and Landolt [10] reviewed the theory and applications of electrochemical microfabrication technology, focusing on electrodeposition and dissolution processes. They discussed the important role of mass transport and current distribution and showed how numerical modeling contributes to the present understanding of critical process parameters. Tan [11] discussed the effects of electrode inhomogeneity and electrochemical heterogeneity on electrode processes.

Techniques that could be used to gain understanding of localized deposition or dissolution processes would allow better deposition rates and dimensional control. In this chapter we provide an overview of laboratory techniques that could be used to study localized electrodeposition and elelctrodissolution processes and their influencing factors. Attention is directed principally to technological innovations that have been made over the past decades, such as the wire beam electrode (WBE) [2,11].

7.1 EXPERIMENTAL METHODS FOR SENSING LOCALIZED ELECTRODEPOSITION AND DISSOLUTION

Traditionally, the common practices of controlling the distribution of electrodeposition or electrodissolution current are performed largely on a trial-and-error basis by adjusting bath parameters, such as the geometry of the bath, the positions of anodes and cathode, and the composition of electrolyte. These industrial practices rely on personnel experience and sometimes on the assistance of a Hull cell test and mathematical modeling [1].

Use of a Hull cell [12] is a classic experimental method for assessing electroplating current distribution and the quality of deposits. The Hull cell is a trapezoidal container that holds 267 mL of solution and replicates the plating bath on a laboratory scale with a shape that makes it possible to place the test panel at an angle to the anode. As a result, the deposit is plated at different current densities that record the character of electroplating at all current densities within the operating range. The cell is used to qualitatively check the condition of an electroplating bath, such as the optimization of current density range and additive concentration, recognition of impurity effects, and indication of macro-throwing power capability [1,13]. The Hull cell has been used widely as a conventional laboratory technique for qualitative characteristics and the assessment of plating baths by determining for a given set of parameters the relationships between the distance of the cathode from the anode and the thickness of the electroplated layer.

However, the oversimplified design of the Hull cell introduces limitations in delineating clearly the distribution of electrodeposition currents. It does not provide real-time information regarding the distribution of electrodeposition current, and it is undoubtedly tedious. Although some new developments have been made to the Hull cell to allow quantitative determination of the throwing power of a bath, a Hull cell–based experiment has difficulties in fully characterizing an electroplating bath where major electrochemical heterogeneities arise from factors such as the complex geometry of the bath, localized chemical environments in the bath, and the metallurgical inhomogeneity of the workpiece [2].

Mathematical modeling has been used widely as a tool to predict current distribution and to facilitate the design of electrochemical cells for electrodeposition and electrodissolution. Various mathematical models have been developed to describe microscopic and macroscopic processes occurring during electroplating processes. Macroscopic electroplating models are designed to predict current densities and plating rates over centimeter-to-meter lengths; mesoscale models are developed to predict interface stability and the shape of dendrites on submicrometer-to-millimeter scales; and atomistic models explicitly track the motion of individual atoms to study fundamental mechanisms of bonding and charge transfer at interfaces. A macroscopic model is used by engineers to position electrodes to make the current density relatively uniform and to determine the highest local plating current density that might lead to a rough deposit. These models are very useful since they can provide theoretical predictions of plating behavior and thus they could be employed to control the distribution of electrodeposition currents. However, in practice, accurate modeling of a practical electrochemical process can be very difficult. This is because in electrochemical systems, boundary conditions are nonlinear and time dependent. There are many complex factors that can affect the actual electroplating current distribution. For example, the development of micrometer-scale dendrites could dramatically change the apparent resistance of an electrode at the interface and thus significantly change the behavior of an electroplating cell. For this reason, experimental measurements are essential for characterizing practical electrochemical processes and should be used as a complement to mathematical models.

The ability to characterize localized deposition and dissolution is particularly important if they are used for fabricating complex three-dimensions micropatterns on metal surfaces. To characterize an electrodeposition or electrodissolution process, local chemical, electrochemical, or morphological information such as local electroplating currents has to be obtained from every location on the electrode surface. Unfortunately, direct measurement of localized quantity such as nonuniform current distribution is not an easy task. This is due primarily to the fact that traditional electrochemical methods such as electrode potential measurement, have been developed based on a uniform electrode model. In principle, they are applicable only to an ideally uniform electrode surface where the electrode potential at any location equals that of the entire electrode surface. If an electrode surface is not uniform, this method only detects a potential that is a mixture or average of contributions from many local potentials, none of which we can evaluate independently.

Various optical and physical methods have been developed and used to gain understanding of localized electrode processes. Video imaging of an electrode surface was used as a means of studying localized copper deposition and dissolution [14]. A localized copper dissolution phenomenon was observed by examination of video images of the copper deposits taken continuously as they grew, and by measurement of the quantity of the deposit and the fractal dimension of fixed interior regions of the deposit as a function of time throughout the deposit process [14]. This phenomenon was attributed to local concentration cells. Synchrotron x-ray microradiography with high time and lateral resolution was used to observe the copper localized electrochemical deposition process in real time [15]. It enabled the accurate control of anode motion, leading to improvement in the quality of the microstructure. Microradiography reveals a deposition mechanism that differs as a function of the distance between the electrode (the anode) and the growing structure (the cathode) [15].

A micro-reference electrode was used to monitor the potential at the localized copper electrochemical deposition site [16]. Measurement of this potential permitted estimation of the local copper ion concentration resulting from their dynamic consumption by electrochemical reduction and their supply by mass transport. The mass-transfer rate of copper ions was estimated using a theoretical calculation based on diffusion. The surface morphology and internal structure were found to be affected significantly by this local copper concentration [16]. Cyclic voltammetry, scanning Auger microscopy, and in situ optical microscopy have been used to study local processes of electrochemical deposition and dissolution of Cu on microstructured Ti substrates. Electrodeposition studies showed that on TiN microareas, bulk Cu deposition starts at a significantly lower cathodic overpotential than on surrounding TiO_2 domains. Deposition on the entire TiN/TiO_2 surface can be achieved by applying cathodic overpotentials higher than 220 mV. At relatively low cathodic overpotentials, the electrodeposition reaction on laser microstructured TiO_2 surfaces was found to occur only at pits and trenches [17].

Electrochemical impedance spectroscopy (EIS) was used to characterize the deposition of nickel from unbuffered acid sulfate electrolytes, from which a wide range of deposit morphologies and current efficiencies are possible [18]. It was reported that the impedance spectra for nickel deposition at 20 mA/cm^2 consisted of one or two characteristic loops whose frequency and capacitance were dependent on the electrowinning conditions and indicative of the resulting deposit morphology. A single high-frequency capacitive loop, on the order of 1 kHz, correlated with good-quality deposits that were flat, smooth, and ductile. The presence of a low-frequency loop, on the order of a few hertz, indicated a degraded deposit morphology which showed localized dark, glassy areas and were cracked, curled, and brittle. The second loop may be associated with a diffusion-controlled component in the reaction mechanism. Deposits of intermediate quality had impedance spectra consisting of both types of loops. The authors claimed a correlation between the impedance spectra and the deposit quality [18].

The advent of variously designed scanning probes has enabled the sensing of nonuniform electrode surfaces. These scanning probe techniques, such as scanning tunneling microscopy (STM), atomic force microscopy (AFM), and scanning electrochemical microscopy (SECM), have provided us with the possibility of efficient detection of local electrochemical current distribution and characterization of the nucleation and growth of electrocrystallization. For example, an ST microscope was used as a nanoelectrode to study localized electrodeposition of metal clusters on the nanometer scale [19]. The width and height of the clusters, which can be grown even with diameters below 10 nm, are determined by the diameter of the STM tip apex, the distance between tip and substrate, the substrate potential, and the amount of Co transferred to the substrate via the tip. The influence of these parameters on cluster growth was understood assuming diffusion as the mechanism of Co transfer from the tip to the substrate [19]. In another experiment, STM was used as a tool for in situ imaging of patterns formed at electrode surfaces and for investigating the possibility of electrodepositing localized (<5 nm) nanostructures on Au by very short (<100 ns) electrochemical pulses [20].

A limitation of scanning probe technologies is that they can scan only a very small electrode area, and it is very difficult to ensure that the scanning tips are positioned correctly over reaction sites of interest. For example, it is very difficult to identify precursor metal clusters over a relatively large electrode surface, and thus successful imaging of localized deposition or dissolution initiation can be challenging in practical experiments.

Over the past decade an electrochemically integrated electrode array, the wire beam electrode (WBE), has been employed as a new method of characterizing nonuniform electrode processes. The WBE has been used as a unique technique for in situ measurement, characterization, and valuation of nonuniform electrodeposition and electrodissolution processes. In typical experiments, the WBE has been shown to be useful in characterizing electroplating, in monitoring anodic polymerization, in sensing anodic dissolution and localized corrosion processes, and in assessing factors affecting electrochemical current distribution [2,21,22].

7.2 SENSING LOCALIZED ELECTRODEPOSITION USING A WBE

It is well known that in electroplating, objects with sharp corners and features tend to have thicker deposits on the outside corners and thinner ones in the recessed areas. This is due to plating current flows more densely to sharp edges than to the less accessible recessed areas (i.e., the current distribution is not uniform). This nonuniform current distribution is not acceptable in many industry applications, in particular in the microelectronics industry, where precise deposition of copper interconnection lines down to less than 0.02 μm width is needed for integrated-circuit fabrication. Indeed, modern electroplating operations require the ability for precise design and control of electrodeposition, which traditionally were achieved through laboratory trial and error.

The WBE method has been used to measure and monitor the electroplating process by mapping the distributions of currents over a WBE surface under cathodic polarization [2,21,22]. The nonuniform distribution of electroplating currents was mapped as an indicator of electrochemical heterogeneity over an electrode surface. In an electrodeposition experiment, the secondary current distribution, rather than the primary current distribution, was found to play a major role in determining the distribution of electroplating currents [21]. The Nonuniform electrodeposition and electrodissolution were characterized by measuring and identifying characteristic patterns in electrodeposition and electrodissolution current distribution maps. Various patterns of electrodeposition current distribution have been obtained from Watts nickel plating and bright acid copper plating baths, with the effects of several factors, such as bath concentration, temperature, agitation, and electrolyte flow. Typical patterns of electrodissolution current distribution have also been detected over a WBE surface under anodic dissolution [22]. A typical laboratory plating bath incorporating a WBE is shown in Figure 7.1.

In a typical electroplating experiment [22], electrodeposition was carried out under potentiostatic control with a WBE as the cathode. This was realized by

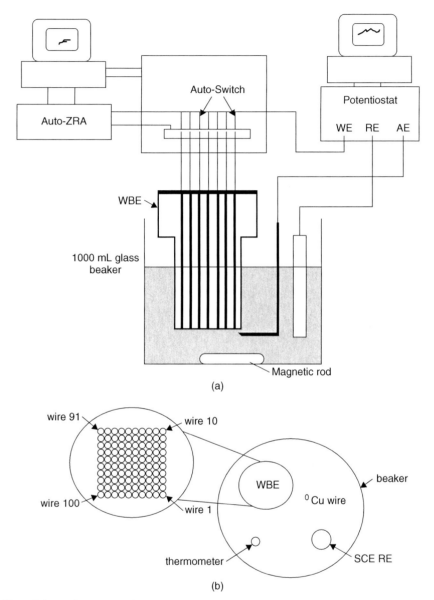

Figure 7.1   (a) General setup for electroplating experiments; (b) plan view of electroplating bath setup. (From [22].)

connecting the WBE to the negative terminal of a potentiostat while connecting an inert or active anode to the positive terminal of the potentiostat. A saturated calomel electrode was used as the reference electrode for potential measurements. During the experiment, the potentiostate was used to polarize the WBE cathodically below its rest potential. The impressed electroplating currents were measured from local areas of the electrode surface by means of wires located at those areas by connecting a zero-resistance ammeter in sequence between the wire terminal chosen and all other terminals shorted together using a computer-controlled automatic switch. This was repeated for all 100 wires so that an electrodeposition current distribution map could be generated at any point of time during the course of electroplating.

Various characteristic current distribution patterns, which indicate different electrodeposition mechanisms, have been obtained from Watts nickel plating and bright acid copper plating baths, with the effects of several factors, such as bath concentration, temperature, agitation, and electrolyte flow. The most common pattern observable in electroplating current distribution maps was of the *secondary current distribution type* (Figure 7.2), a pattern that shows higher currents along the electrode edges, with the magnitude decreasing in a contour-like manner toward the center of the WBE surface [2,22]. This result is in agreement with the well-known phenomenon that electroplating currents tend to concentrate at edges and corners of a workpiece, due to nonuniform concentration distribution over the WBE surface. The edges and corners of the WBE surface have a more sufficient supply of metal ions as well as of hydrogen ions and oxygen, through two- or three-dimensional diffusion available at these areas, to support cathode reactions.

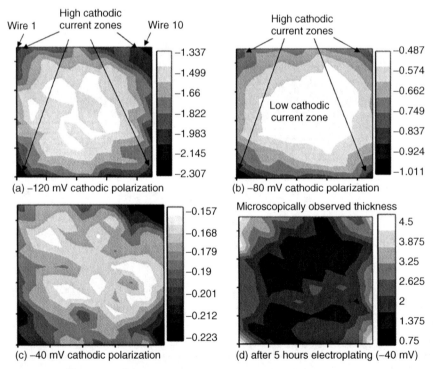

Figure 7.2 Current distribution maps recorded from a Watts nickel bath after electroplating for 30 minutes and the thickness profile of nickel deposit on a WBE after 5 hours' plating with −40 mV of cathodic polarization voltage (in μm, determined using a microscope). (From [22].)

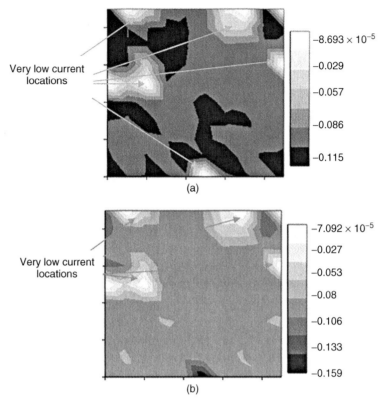

Figure 7.3 Current distribution maps recorded from a Watts nickel bath after (a) 30 minutes and (b) 60 minutes of plating at room temperature. Current values are in mA/cm.^{2} (From [22].)

An interesting result obtained from nickel electroplating at ambient temperature is shown in Figure 7.3, which indicates that some areas over the WBE surface had registered only negligible electroplating currents, indicating the poor covering power of the bath. The covering power is the ability of a plating solution to produce a relatively uniform distribution of metal on a cathode of irregular shape. When the WBE surface was inspected visually after rinsing with deionized water and dried after electroplating, such areas correlate well with wires whose surfaces suffered either the partial or total absence of nickel deposition. This result suggests that temperature affected covering power significantly.

The influences of several electroplating parameters, including the geometry of the bath, the concentration and movement of the electrolyte, and the bath temperature, on the uniformity of electroplated coatings have also been investigated [2,21,22]. When bath concentration decreased, the electrodeposition current pattern was found to change into the *primary current distribution type*, a pattern that shows higher currents at locations in proximity to the anode and lower at regions farther away. When nonuniform fluid movement was introduced to baths, current distribution was found to change into a flow-controlled pattern that shows higher plating currents at locations where the flow rate was higher [22].

A characteristic current distribution pattern showing the peeling-off of a non-coherent deposit (Figure 7.4) was also detected from bright acid copper plating baths where copper deposits were washed away from electrode surfaces. Lower offset voltages, a suitable amount of bath agitation, elevated temperatures, appropriate cathode–anode separation, and bath concentration were found to contribute to the enhancement of current distribution uniformity. This work suggests that the WBE is a practical tool for monitoring, characterizing, and optimizing electrodeposition and electrodissolution processes. The WBE could also be used to verify the accuracy and completeness of mathematical modeling of electrodeposition and electrodissolution processes.

7.3 SENSING LOCALIZED ELECTRODISSOLUTION USING A WBE

Electropolishing and electromachining are examples of industry applications of desirable anodic dissolution processes. Electropolishing is a procedure in which anodic dissolution is used to dissolve microelevations selectively on a metal surface to create a smooth surface and in some instances, a mirror finish. This is achieved by connecting the specimen to be polished to the positive dc supply, where it becomes an anode of an electrolytic cell. This process is often referred to as *reverse plating*. Electropolishing has numerous applications in the laboratory and in industry for preparing scratch-free surfaces for microscopic examination, for removing edge burrs or deformations produced by a mechanical cutting tool, and for improving surface appearance and reflectivity. Electromachining is also based on controlled anodic dissolution of a metal workpiece with a preshaped cathode "tool" in an electrolytic cell. At the area of the anodic workpiece surface selected, metal is dissolved, and thus the tool shape is copied into the workpiece. The electromachining technique is used widely in duplicating, drilling, and sinking operations for the generation of passages, cavities, holes, and slots in metal parts such as dies, molds, and turbine blades. For both electropolishing and electromachining, the key is to achieve controlled and selective dissolution of metal surfaces. In electropolishing, anodic current has to be concentrated on the microelevations of the specimen surface to achieve microleveling of the surface. However, electropolishing is possible only if parameters such as the applied voltage and electrolyte composition are selected properly; otherwise, there could be a different anodic dissolution mechanism leading to etching or pitting of the specimen surface. Obviously, the distribution of anodic dissolution current is critical for successful electropolishing and electromachining operations.

The WBE has been utilized to characterize the anodic dissolution current and affecting factors [22]. After the nickel electroplating processes shown in Figure 7.4 was completed, the polarization was changed from cathodic to anodic, and thus the WBE surface was under anodic dissolution (in the same bath). Figure 7.5 shows typical anodic current distribution maps recorded from the nickel-plated WBE. As

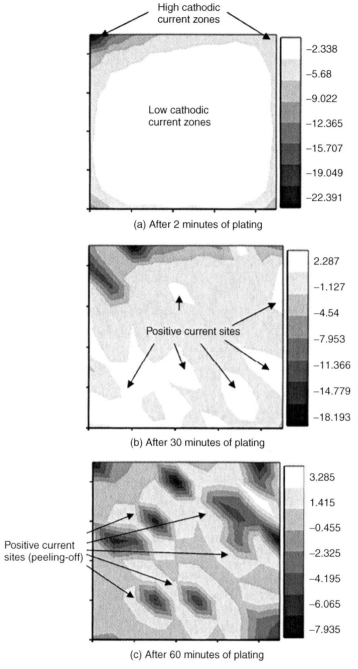

(a) After 2 minutes of plating

(b) After 30 minutes of plating

(c) After 60 minutes of plating

Figure 7.4   Current distribution maps of plating with a bright acid copper bath (-500 mV cathodic polarization, 1% concentration, 20 to 21 °C, 200 rpm bath agitation, anode 5 mm from wire 1). Current values are in mA/cm^2. (From [22].)

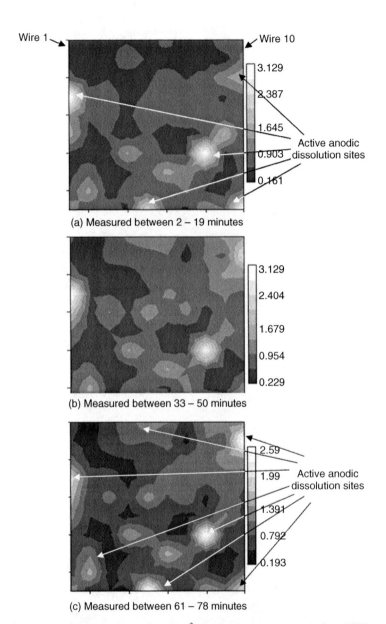

Figure 7.5   Electrodissolution current (in mA/cm^2) distribution maps measured after a WBE was under anodic electrodissolution for various periods under anodic dissolution (+40 mV). A positive current indicates an anodic reaction. Current values are in mA/cm^2. (From [22].)

shown in Figure 7.5, active electrodissolution spots were clearly observable. The pattern of nonuniform electrodissolution current distribution appeared to indicate etching of the WBE surface, with active dissolution at spots that appeared to be distributed randomly over the surface. The active dissolution area appeared to be growing with extension of the electrodissolution period [22].

Localized anodic dissolution is also a critical process in localized corrosion. Traditionally, the most common method of studying the behavior of localized anodic dissolution is through measurement of the characteristic pitting potential using potentiostatic, quasi-potentiostatic, potentiodynamic, and galvanostatic techniques, complemented by ex situ surface analytical techniques such as SEM and EDS [23]. It is believed that the anodic passivity often breaks down at potentials higher than the pitting potential [23]. Impedance spectroscopy and other electrochemical techniques have been used to study electrodissolution and pitting characteristics [24]. In typical experiments on a 2024A1-T4 alloy, potentiodynamic measurements from −1.0 to 1.5 V show that above 0.0 V the electrodissolution rate increased in the order saturated $AlCl_3 < I^- < Br^- < Cl^-$. Anodic potential step experiments above the pitting potential of 2024A1-T4 in Cl^-, Br^-, and I^- for a duration of 15 minutes showed that pits were initiated at the periphery of the disk and contained black films; the number of pits increased as the anodic potential increased. The impedance spectra of 2024A1-T4 at anodic potentials above the pitting potential indicate two capacitive loops in the complex plane plots and two maxima at high frequencies in the corresponding phase-angle Bode plots when pits were clearly visible [24].

However, the electrochemical methods used in these experiments would not provide spatial and temporal information on local anodic dissolution processes, and thus have to rely on ex situ surface analytical techniques such as SEM for the identification of pits. To overcome this limitation, scanning probes such as scanning electrochemical microscopy (SECM) have been used to detect the redox species involved in a corrosion process. In a typical experiment, SECM was used to detect Fe^{2+} ions originating from the electrodissolution of iron at the anodic sites on the metal surface during the immersion of pure iron in a neutral aqueous solution [25].

In a different approach, scanning probes are used in conjunction with the WBE method. Figure 7.6 shows an experimental setup using a WBE, in conjunction with the scanning reference electrode technique (SRET), to characterize the anodic dissolution behavior of aluminum AA1100 [26]. The mapping of anodic current was carried out in a three-electrode electrochemical cell with an aluminum WBE as the working electrode, a saturated calomel electrode (SCE) as the reference electrode, and a platinum electrode as the auxiliary electrode. A SRET instrument functioned as an electrochemical cell containing a 0.5 M NaCl-based solution. Sodium dichromate was added to the base solution for studying inhibition effects. The WBE surface was polarized anodically under potentiostatic control, which was realized by connecting the WBE to the WE terminal of a potentiostat. The impressed anodic polarization current for each wire was measured by connecting a zero-resistance ammeter in sequence between the wire terminal chosen and all

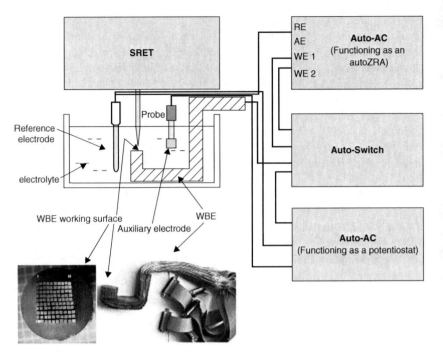

Figure 7.6   Experiment to study anodic dissolution using a WBE in combination with SRET, and photographs showing the WBE and its working surface. (From [26].)

other terminals (shorted together) using a computer-controlled automatic switch. This was repeated for all 100 wires so that an anodic polarization current map representing the current distribution across the WBE surface was generated. SRET measurements were also carried out, alternately with WBE measurements, over the anodically polarized WBE surface, to map current distribution in the electrolytic phase.

Figure 7.7 shows typical maps measured from a WBE surface anodically polarized to (b) approximately 7 mV below and (c) 13 mV above the pitting potential of the metal. As shown in the WBE map of Figure 7.7c, anodic dissolution currents increased significantly when the WBE was polarized to a potential above the pitting potential, and the currents were concentrated on a few major anodic sites with a maximum dissolution current value of 0.718 mA/cm^2. SRET measurements were also able to detect anodic dissolution processes, revealing anodic dissolution behavior similar to that in the corresponding WBE map. However, Figure 7.7a and b show that the SRET was not sensitive enough to monitor the free corrosion process occurring on the surface of aluminum exposed to a relatively mild corrosion environment or under low anodic polarization voltage.

When corrosion inhibitor potassium dichromate was added to the corrosive solution, as shown in Figure 7.8, the SERT measurements were unable to detect ionic currents in the solution even when the WBE was polarized to -611 mV,

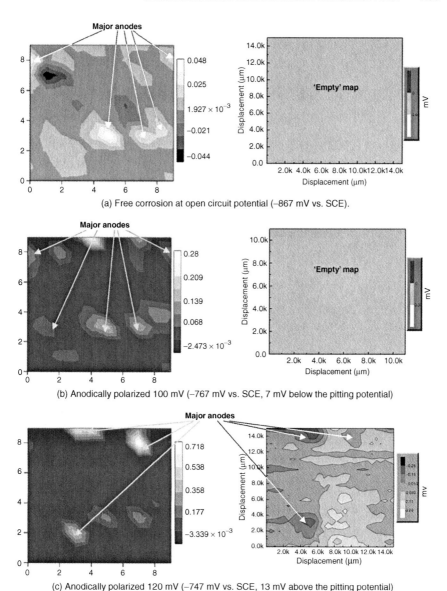

(a) Free corrosion at open circuit potential (−867 mV vs. SCE).

(b) Anodically polarized 100 mV (−767 mV vs. SCE, 7 mV below the pitting potential)

(c) Anodically polarized 120 mV (−747 mV vs. SCE, 13 mV above the pitting potential)

Figure 7.7 WBE and SRET maps measured in an aluminum WBE exposed to 0.5 M NaCl solution before and after applying external anodic polarization. (From [26].)

39 mV above its pitting potential. This result clearly demonstrates that potassium dichromate effectively inhibited anodic dissolution. However, when polarization was maintained and extended to 30 minutes, as shown in Figure 7.8b, major changes occurred in the WBE and SRET maps. Major anodic dissolution sites, which are different from the major anodic sites in Figure 7.8a, were formed during this period

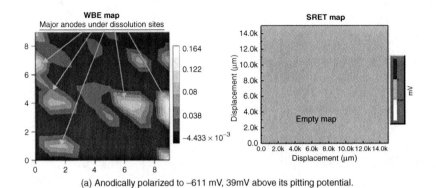

(a) Anodically polarized to –611 mV, 39mV above its pitting potential.

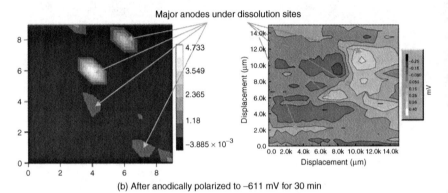

(b) After anodically polarized to –611 mV for 30 min

Figure 7.8 WBE and SRET maps measured from an aluminum electrode exposed to a solution containing 0.5 M NaCl and 0.5 M $K_2Cr_2O_7$ (a) immediately and (b) 30 minutes after applying external anodic polarization. (From [26].)

and are observable in both WBE and SRET maps. Anodic dissolution currents increased significantly, from 0.164 mA/cm^2 to a maximum value of 4.733 mA/cm^2. SRET measurements were able to detect anodic dissolution processes, as shown in Figure 7.8b, revealing anodic dissolution behavior similar to that shown in the corresponding WBE map. This result suggests that with inhibitors, larger polarizations were required for pitting to be initiated; in particular, a long duration was required for a pit to grow [26].

These experiments confirm the applicability of the WBE and SRET methods for sensing electrochemical currents from surfaces under anodic polarization with and without the effects of corrosion inhibitors. Under polarization conditions, SRET and WBE techniques were used successfully in determining the corrosion behavior and the profile below and above the pitting potential. The WBE method appeared to be more sensitive and could provide more detailed information on corrosion processes than could the SRET methods. This work has also revealed two different mechanisms of localized corrosion initiation: (1) the initiation of localized corrosion due to the disappearance of minor anodes; and (2) the initiation of localized corrosion due to the initiation of new localized anodic sites [26].

7.4 SENSING NONUNIFORM ELECTROCHEMICAL DEPOSITION OF ORGANIC COATINGS

Cathodic and anodic polarization are methods of promoting the growth of organic or inorganic coating of special properties on metal surfaces [13]. Cathodic electrodeposition of organic coatings such as epoxy is the most common and widely used coating process for providing the first prime coat to a variety of industrial products, such as automobiles. The advantages of cathodic electrodeposition include its automated operation, the high level of paint utilization, the low level of pollution and high throwing power to coat recessed areas of complex metal shapes, and avoidance of the electrochemical dissolution of metal. Fundamental aspects of cathodic electrodeposition process have been reported by Beck [27].

Anodic electrodeposition is also useful; for example, anodizing is used to grow oxide films on certain metals, such as on an aluminum surface to form a thin barrier oxide film or a thick oxide coating with a high density of microscopic pores. These films or coatings have diverse and important applications, including architectural finishes, prevention of corrosion, and electrical insulation. In an anodizing cell, the aluminum workpiece is made the anode by connecting it to the positive terminal of a dc power supply. The bath composition and the anodic polarization voltage are primary determinants of whether the film will be barrier or porous. Anodic polarization is also a method of electrodeposition of protective polymer coatings such as anticorrosion polyaniline (PANI) coatings on metallic surfaces. Compared to chemical polymerization of PANI coatings, electrochemical deposition is believed to be a more efficient way of controlling the chemical and physical properties, such as adhesion, molecular weight, and the redox state of a polymer coating.

An important requirement for PANI film electrodeposition is to ensure that the deposit has uniform thickness and thus is protective. The WBE has been employed to characterize and optimize anodic electrodeposition of PANI coatings and also to evaluate the anticorrosion performance of electrodeposited coatings [28,29]. The electrodeposition of PANI was carried out by anodically polarizing a WBE surface and measuring the distribution of localized anodic polarization currents from every location. Figure 7.9 shows an experimental setup used for anodic electrodeposition operations.

Electrodeposition of PANI was performed in a three-electrode electrochemical cell with a WBE as the working electrode, a saturated calomel electrode (SCE) as the reference electrode, and a high-purity graphite rod as the auxiliary electrode. The impressed anodic currents were measured from local areas of the electrode surface by means of wires located at those areas by connecting a zero-resistance ammeter in sequence between the wire terminal chosen and all other terminals shorted together using a computer-controlled automatic switch. This was repeated for all 100 wires so that an anodic current distribution map could be generated at any point in time during the course of anodic electrodeposition of PANI coating. Figure 7.10 shows a typical anodic current map measured from an aluminum AA1100 WBE surface during electrodeposition of PANI in 1.0 M tosylic acid

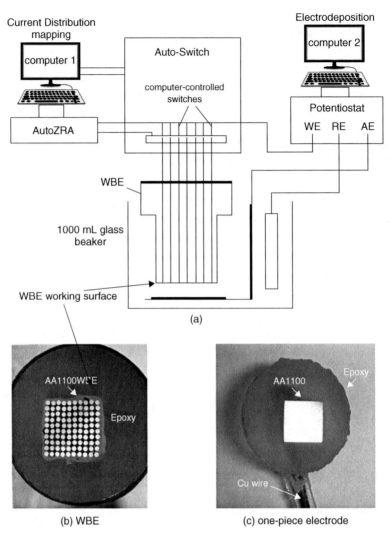

Figure 7.9 Experimental setup for PANI electrodeposition on a WBE and photographs of an AA1100 WBE and a one-piece electrode. (From [28, 29].)

containing 0.16 M aniline under constant anodic polarization (1.25 V vs. SCE). Under these experimental conditions (Figure 7.10), the distribution of anodic current and the actual electrodeposited PANI film were nonuniform. It is evident that higher anodic polarization current was distributed nonuniformly over the upper region of the WBE surface. This current distribution pattern corresponds well with the area that was more PANI deposits. The anodic current density on the lower half of the WBE surface is three or four orders of magnitude lower, which agrees with the fact that little PANI deposit is visible in these regions. This result shows clearly that the anodic current maps are correlated with the electrodeposition behavior of

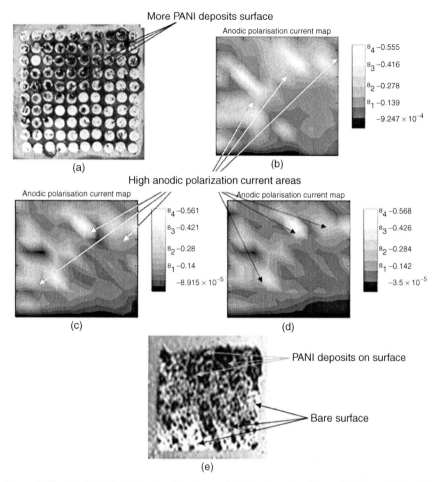

Figure 7.10 (a) AA1100 WBE after 60 minutes of PANI electrodeposition; (b), (c), and (d) anodic polarization current maps after 30, 40, and 50 minutes of PANI electrodeposition, respectively; (e) one-piece AA1100 electrode surface with PANI electrodeposited under the same conditions. (From [28].)

PANI deposits and thus can be used as a means of monitoring and studying the electrodeposition of PANI coating.

Surface pretreatment of the aluminum surface was applied by cathodic polarization to remove an oxide layer naturally formed on an AA1100 electrode. After pretreatment, as shown in Figure 7.11, PANI coating was deposited successfully on an AA1100 electrode. The anodic polarization current values shown in Figure 7.11 are much larger (maximum approximately 3.0 mA/cm^2) than those shown in Figure 7.10 (maximum approximately 0.56 mA/cm^2), suggesting that the cathodic pretreatment has significantly improved the efficiency of PANI electrodeposition on an AA1100 surface. It is also interesting to note that higher

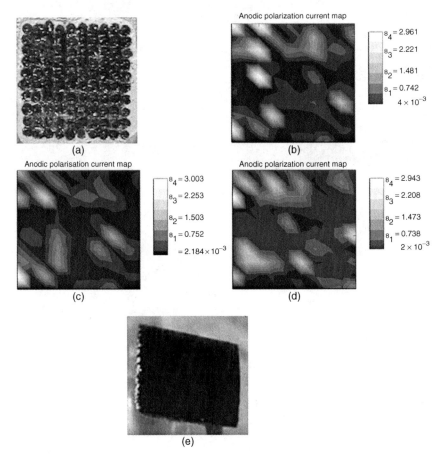

Figure 7.11 (a) AA1100 WBE after pretreatment and 35 minutes of PANI electrodeposition; (b), (c), and (d) anodic polarization current maps after 20, 25, and 30 minutes of PANI electrodeposition; (e) one-piece AA1100 electrode surface with PANI electrodeposited under the same conditions. (From [29].)

current regions, as indicated by the brighter spots in Figure 7.11, were distributed randomly over the WBE surface and their locations shifted frequently, leading to a more uniform distribution of PANI deposits over the WBE surface. Random distribution of anodic polarization current could be a characteristic pattern for producing PANI coating that covers an electrode completely. It should be noted that according to the anodic polymerization mechanism, the local anodic current density may not be proportional to the distribution of a PANI deposit over an anodically polarized electrode surface, unlike in electroplating metal coatings, where the local cathodic current distribution determines the deposit thickness. However, it is expected that local anodic polarization current density should affect the radical cation reactions and thus determine the localized growth rate and the characteristic of PANI deposits [29].

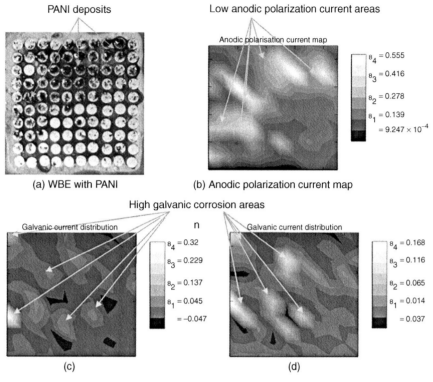

Figure 7.12 (a) AA1100 WBE with PANI deposits; (b) an anodic polarization current map measured after 20 minutes of PANI electrodeposition; (c) and (d) galvanic corrosion current maps showing galvanic current distribution after 5 and 10 minutes of immersion in 0.5 M NaCl solution. (From [29].)

After PANI electrodeposition was completed, the PANI-coated AA1100 WBEs were exposed to an aerated 0.5 M NaCl solution to test under free corrosion conditions. Figure 7.12 shows galvanic current maps measured from a nonuniform PANI-deposited electrode to study the relationship between the PANI deposits and their corrosion performance. If we carefully examine and compare the photograph and the anodic polarization current map shown in Figure 7.12a and b with the galvanic current maps shown in Figure 7.12c and d, it can be seen that the cathodic areas are mainly locations with more deposits, suggesting that PANI deposits behaved as cathodes over the WBE surface. The major anodic areas on the WBE are located primarily in areas with fewer PANI deposits; for example, the anodes shown in Figure 7.12c and d are located at neighboring wires next to the cathodes. The anodic current values decreased with the extension of electrode exposure, with the maximum anodic current value decreasing from 0.321 mA/cm^2 to 0.168 mA/cm^2. This indicates that an aluminum electrode with a PANI deposit underwent a passivation process, with the PANI behaving as an effective cathode [29].

These experiments indicate that the WBE is a practical tool for monitoring, characterizing, and optimizing electrodeposition processes and for evaluating the anticorrosion performance of electrodeposited coatings.

REFERENCES

1. F. A. Lowenheim, *Electroplating*, McGraw-Hill, New York, 1978.

2. Y. J. Tan and K. Y. Lim, Understanding and improving the uniformity of electrodeposition, *Surface and Coating Technology*, 167 (2003), 255–262.

3. S. H. Yeo, J. H. Choo, and K. S. Yip, Localized electrochemical deposition: the growth behavior of nickel micro-columns, in *Micromachining and Microfabrication Process Technology*, J. M. Karam and J. Yasaitis, Eds, Proceedings of the Society of Photo-optical Instrumentation Engineers, Vol. 4174, 2000, pp. 30–39.

4. J. D. Madden and I. W. Hunter, Three-dimensional microfabrication by localized electrochemical deposition, *Journal of Microelectromechanical Systems*, 5 (1996), 24–32.

5. A. Jansson, G. Thornell, and S. Johansson, High resolution 3D microstructures made by localized electrodeposition of nickel, *Journal of the Electrochemical Society*, 147 (2000), 1810–1817.

6. E. M. El-Giar, R. A. Said, G. E. Bridges, and D. J. Thomson, Localized electrochemical deposition of copper microstructures, *Journal of the Electrochemical Society*, 147 (2000), 586–591.

7. M. D. Perez, E. Otal, S. A. Bilmes, G. J. A. A. Soler-Illia, E. L. Crepaldi, D. Grosso, and C. Sanchez, Growth of gold nanoparticle arrays in TiO_2 mesoporous matrixes, *Langmuir*, 20 (2004), 6879–6886.

8. L. Q. Wu, K. Lee, X. Wang, D. S. English, W. Losert, and G. F. Payne, Chitosan-mediated and spatially selective electrodeposition of nanoscale particles, *Langmuir*, 21 (2005), 3641–3646.

9. L. Oniciu and L. Muresan, Some fundamental aspects of leveling and brightening in metal electrodeposition, *Journal of Applied Electrochemistry*, 21 (1991), 565–574.

10. M. Datta and D. Landolt, Fundamental aspects and applications of electrochemical microfabrication, *Electrochimica Acta*, 45 (2000), 2535–2558.

11. Y. J. Tan, Sensing electrode inhomogeneity and electrochemical heterogeneity using an electrochemically integrated multi-electrode array, *Journal of the Electrochemical Society*, 156 (2009), C195–C208.

12. R. O. Hull, *American Electroplating Society*, 27 (1939), 52.

13. L. J. Durney, in *Electroplating Engineering Handbook*, 4th ed., Van Nostrand Reinhold, New York, 1984, p. 461.

14. Y. Deng and M. Pritzker, Experimental evidence for local dissolution during galvanostatic copper electrodeposition, *Journal of Electroanalytical Chemistry*, 336 (1992), 25–34.

15. S. K. Seol, A. R. Pyun, Y. K. Hwu, G. Margaritondo, and J. H. Je, Localized electrochemical deposition of copper monitored using real-time x-ray microradiography, *Advanced Functional Materials*, 15 (2005), 934–937.

16. J. C. Lin, T. K. Chang, J. H. Yang, Y. S. Chen, and C. L. Chuang, Localized electrochemical deposition of micrometer copper columns by pulse plating, *Electrochimica Acta*, 55 (2010), 1888–1894.

17. O. Voigt, B. Davepon, G. Staikov, and J. W. Schultze, Localized electrochemical deposition and dissolution of Cu on microstructured Ti surfaces, *Electrochimica Acta*, 44 (1999), 3731–3741.

18. M. Holm and T. J. O'Keefe, Evaluation of nickel deposition by electrochemical impedance spectroscopy, *Journal of Applied Electrochemistry*, 30 (2000), 1125–1132.

19. W. Schindler, D. Hofmann, and J. Kirschner, Localized electrodeposition using a scanning tunneling microscope tip as a nanoelectrode, *Journal of the Electrochemical Society*, 148 (2001), C124–C130.

20. G. Ertl, Pattern formation at electrode surfaces, *Electrochimica Acta*, 43 (1998), 2743–2750.

21. Y. J. Tan, Studying non-uniform electrodeposition using the wire beam electrode method, *International Journal of Modern Physics B*, 16 (2002), 144–150.

22. Y. J. Tan and K. Y. Lim, Characterising nonuniform electrodeposition and electrodissolution using the novel wire beam electrode method, *Journal of Applied Electrochemistry*, 34 (2004), 1093–1100.

23. M. Metikoshukovic and I. Milosek, Electrochemical methods in the study of localized corrosion attack, *Journal of Applied Electrochemistry*, 22 (1992), 448–455.

24. L. Chen, N. Myung, P. T. A. Sumodjo, and K. Nobe, A comparative electrodissolution and localized corrosion study of 2024Al in halide media, *Electrochimica Acta*, 44 (1999), 2751–2764.

25. A. C. Bastos, A. M. Simoes, S. Gonzalez, Y. Gonzalez-Garcia, and R. M. Souto, Imaging concentration profiles of redox-active species in open-circuit corrosion processes with the scanning electrochemical microscope, *Electrochemistry Communications*, 6 (2004), 1212–1215.

26. T. Liu, Y. J. Tan, B. Z. M. Lin, and N. N. Aung, Novel corrosion experiments using the wire beam electrode: IV. Studying localized anodic dissolution of aluminum, *Corrosion Science*, 48 (2006), 67–78.

27. F. Beck, in *Comprehensive Treatise of Electrochemistry*, J. O'M. Bockris, B. E. Conway, E. Yeager, and R. E. White, Eds., Vol. 2, Plenum Press, New York, 1981, p. 537.

28. T. Wang and Y. J. Tan, Electrodeposition of polyaniline on aluminum alloys for corrosion prevention: a study using the wire beam electrode (WBE), *Materials Science and Engineering B*, 132 (2006), 48–53.

29. T. Wang and Y. J. Tan, Understanding electrodeposition of polyaniline coatings for corrosion prevention applications using the wire beam electrode method, *Corrosion Science*, 48 (2006), 2274–2290.

8

Versatile Heterogeneous Electrode Processes

Electrochemical heterogeneity is a versatile natural phenomenon. It is not only the root cause of localized corrosion, nonuniform electrodeposition, and uneven electrodissolution; it also plays critical roles in many scientific phenomena and has found applications in a diverse range of engineering devices. For example, electrochemical heterogeneity has been reported to exist in biological systems, such as on bacterial cell surfaces [1–3] and to involve the sources of electrical activity in the brain [4]. In biological processes, an electrochemical potential gradient in a cell determines the direction in which an ion moves by diffusion or active transport across a membrane. This electrochemical potential energy is used for the synthesis of adenosrine triphosphate, which transports chemical energy within cells for metabolism by oxidative phosphorylation [1].

Electrochemical heterogeneity also plays an important role in electrochemical energy devices such as fuel cells. The nonuniformity of current distribution within a fuel cell is critical to maintaining optimal performance of the fuel cell stack, minimizing polarization loss, and optimizing water management. For example, the current density distribution in a planar solid oxide fuel cell unit with cross-flow configuration is not uniform [5]. In polymer electrolyte membrane fuel cells, electrochemical heterogeneity can cause cathode flooding, a problem that could determine overall fuel cell performance. Determination of the current density distribution is critical to understanding key phenomena in fuel cell systems, including water management. Mench and Wang described a method for the determination of current density distribution in an operating polymer electrolyte membrane fuel cell by taking multiple current measurements of an electrode array simultaneously, allowing transient distribution detection with a multichannel potentiostat [6]. Their

Heterogeneous Electrode Processes and Localized Corrosion, First Edition. Yongjun Tan.
© 2013 John Wiley & Sons, Inc. Published 2013 by John Wiley & Sons, Inc.

technique could contribute to the knowledge and understanding of key phenomena, including water management and species distribution in fuel cells, and could be used for predicing cathode flooding. Hwang et al. [7] mapped local current distribution on a fuel cell experimentally using a specially-designed single-cell fixture with an array of 16 individual conductive segments on the composite plate. The effects of flow-field patterns, dew points of the cathodic feedings, and cathodic stoichiometrics on the local current distribution were examined. The transient variation of the local current distribution on the cathode under supersaturated conditions was visualized further to illustrate the flooding phenomena in different flow patterns [7]. Indeed, knowledge and understanding of the local current distributions in fuel cell systems are critical for effective design of the fuel cell components.

Electrochemical heterogeneity is a factor affecting the performance of cathodic protection. It has been reported that for various reasons cathodic protection systems at compressor stations, refineries, and other industrial plants can experience problems due to changes in the cathodic potential distributions [8]. Some areas become underprotected and others overprotected. The effect of localized potentials is even more significant for cathodic protection of steel in the crevice area and under disbonded coating [9]. It has been demonstrated that in the early stage of the corrosion of steel, cathodic protection cannot reach the crevice bottom to protect steel from corrosion, due to the geometric limitations. Corrosion of steel occurs preferentially inside a crevice due to a separation of anodic and cathodic reactions with the depletion of dissolved oxygen in the crevice solution. The main role of cathodic protection in the mitigation of sequential corrosion in steel in a crevice under disbonded coating is to enhance the local solution alkalinity. With the increase in distance from the open holiday, a high cathodic polarization is required to achieve an appropriate cathodic protection level at a crevice bottom. A potential difference always exists between the open holiday area and inside of the crevice, reducing the cathodic protection effectiveness [9]. Chin and Sabde measured current and potential distributions inside the crevice of a simulated disbonded coating with a holiday during cathodic protection [10]. Microelectrodes mounted on the steel surface were used to determine the local current densities inside the crevice. A change in the concentration of dissolved oxygen inside the crevice was monitored with a micro-oxygen electrode. The current distribution was found to be extremely nonuniform. Depending on NaCl concentrations, the local current density on steel exhibited a 100-fold decrease from that at the holiday opening 20 to 40 mm into the crevice. Current and potential distribution curves became less uniform with a decrease in NaCl concentration and solution conductivity. In 0.6 M NaCl, the steady-state local current density on steel at the sealed end of the crevice was 200 times smaller than that at the holiday opening. In 0.001 M NaCl, the local current density at the sealed end became 1000 times smaller than that at the holiday opening. The dissolved oxygen within the crevice was consumed largely by local corrosion cells when the crevice was formed initially [10].

The ability to control electrochemical heterogeneity is a key to successful application of electroforming technology. Methods employed to improve the uniformity of deposition distribution include the conformal anode, cathode shield, and high-frequency pulse current processes. These methods were found to improve the deposition distribution patterns, which are similar to the current density distribution on the cathode [11]. The nonuniformity issue is more serious in cases of electroforming of complicated shapes. In this process a considerable amount of metal is deposited at the corners and edges, and hence the wastage of metal is on the order of 100 to 200% over and above the requirement, due to the inherent nature of the electrodeposition process: namely, nonuniform current and metal distribution [12]. Modifications of the deposit distribution pattern can be achieved by changing the geometry of the part by using auxiliarly anodes or shields or by changing the rack design. Good throwing power is that property of the plating solution that produces a relatively uniform distribution of metal on a cathode of irregular shape. It has been concluded that metal distribution is more uniform when metal ions are injected through narrow slots, due to scattering of the electric field in the deposition cell [12].

Heterogeneous electrode processes have also been utilized in various engineering processes and devices. For example, electrochemical heterogeneity has been utilized in the electrochemical impregnation of nickel hydroxide in the type of porous nickel plaque [13]. Nickel hydroxide is one of the most commonly used active materials for positive electrodes in secondary batteries. Electrochemical impregnation of porous nickel plaques from a nickel nitrate bath was found to produce superior battery electrodes. It is important to ensure the uniformity of current density for quality-control during electrochemical impregnation [13].

Heterogeneous electrochemical dissolution has also been employed in electrochemical micromachining to achieve maximum etching localization [14]. Methods of using a macroscopically nonuniform rotating disk electrode, sprayer flow, and an electrode placed into a cell with chaotic bulk electrolyte mixing were used to achieve control of local rates of electrochemical dissolution. Compared to primary current distribution, localization enhancement takes place in the case of a turbulent flow under hydrodynamic conditions where the removal of dissolution products from the undercutting region is hindered. These conditions (electrochemical reaction limited by the ion mass transport rate, high resistance to the mass transport in the undercutting region) are necessary for localization enhancement [14].

Electrochemical heterogeneity is observed most frequently on metal surfaces, especially on weld joint areas [15]. It also exists on other material surfaces, such as glassy carbon and highly ordered pyrolytic graphite [16,17]. Spatial heterogeneity on carbon surfaces was found to result in nonuniform adsorption at disordered regions and in particular defects [17].

During the past decade, due primarily to the advent of advanced physical and electrochemical techniques, there has been a marked increase in research aimed at understanding various heterogeneous electrode processes. Recent advances in research methodologies have enabled better understanding of nonuniform electrode–electrolyte interfaces; however, electrochemical heterogeneity is still one of the less well understood properties in electrochemical science and engineering. In this chapter we present an overview of techniques that have been utilized for examining various forms of heterogeneous electrochemical processes of scientific and industrial significance. Particular focus is on techniques developed based on the wire beam electrode (WBE) concept and complementary technologies for visualizing and understanding selected phenomena rooted in electrochemical heterogeneity [18,19]. Typical experiments are presented to illustrate the effects of electrochemical heterogeneity on the initiation of electrochemical noise and the use of electrochemical heterogeneity as an energy source.

8.1 SCANNING AND MODELING VARIOUS HETEROGENEOUS ELECTRODE PROCESSES

Mathematical modeling is probably the most widely used tool for understanding and predicting the distribution of electrochemical reaction currents in various systems. Various mathematical models have been developed to describe various microscopic and macroscopic heterogeneous electrochemical processes [5,7,10,13]. Macroscopic models are designed to predict current densities over centimeter-to-meter lengths; mesoscale models are developed to predict interface stability and the shape of dendrites at submicrometer-to-millimeter scales; and atomistic models explicitly track the motion of individual atoms to study fundamental mechanisms of bonding and charge transfer at interfaces. For example, a cellular automaton model that simulates the mesoscopic-scale behavior of an electrode is used to describe an electrode under free corrosion and under polarization [20]. This mesoscopic approach discloses features that are not considered in a standard macroscopic description. First, the metal surface becomes morphologically heterogeneous. Second, the degree of roughness is linked to a specific distribution of active dissolution sites, and the resulting interplay between morphology and kinetics carries with it a local segregation of reactants and the detachment of clusters of varying size [20]. These models are very useful since they can provide theoretical predictions of electrochemical behavior, and thus they could be used to control the distribution of electrodeposition currents. However, in practice, accurate modeling of a practical electrochemical process can be very difficult. This is because in electrochemical systems, boundary conditions are nonlinear and time dependent. There are many complex factors that can affect actual electroplating current distribution, so experimental measurements are essential for characterizing practical electrochemical processes and should be applied as a complement to mathematical models.

In recent years, various scanning probe techniques have been developed and employed to characterize heterogeneous electrochemical processes. Atomic force microscopy (AFM) has been found to possess the lateral resolution needed to detect nano- to micrometer heterogeneous regions on electrode surfaces. The scanning Kelvin probe (SKP) has been used as a noninvasive no-contact vibrating capacitor technique that is capable of mapping Volta potential differences over an electrode surface. SKP and AFM have be employed in combination as scanning Kelvin probe force microscopy (SKPFM), which has been shown to be an in situ technique for gaining a more detailed understanding of localized electrode processes in the microscopic and submicroscopic ranges. Other scanning probe techniques include the scanning vibrating electrode technique (SVET), local electrochemical impedance spectroscopy (LEIS), and scanning electrochemical microscopy (SECM). During recent years, LEIS and SECM are probably the most frequently used scanning probe techniques for probing various heterogeneous electrode processes.

LEIS is a techniques that has also been used for the detection and mapping of defects and local corrosion events in organic coatings [21,22]. The technique has been used to study heterogeneities in organic coatings on steel surfaces due to underfilm deposits and microblisters [21,22]. Various types of intentional local heterogeneities, including chemical defects within the coating, such as absorbed oil, and physical defects, such as subsurface bubbles, underfilm salt deposits, pinholes, and underfilm corrosion, were detected successfully [21]. Experimental results indicate that LEIS can be used to detect the onset of blistering and that it can give a measure of the rate of water diffusion into the organic coating on a microscopic scale. LEIS also gives valuable information concerning the initiation of corrosion at a defect in an organic coating [22]. LEIS is considered to be a powerful tool for the exploration of electrode heterogeneity [23]. Experimental results for the reduction of ferricyanide show the correspondence between local and global impedances [23]. LEIS has also been employed in conjunction with microcapillary techniques and numerical analysis to quantify the passive properties of resulfurized stainless steel after immersion in chloride media [24]. Local electrochemical impedance spectroscopy measurements provided data describing the behavior of the affected matrix at the microscale. For example, the value of the charge transfer and migration of point defects resistance decreases from $51,700 \ \Omega \cdot cm^2$ in sites free of any metallurgical heterogeneity, down to $12,200 \ \Omega \cdot cm^2$ in sites containing a high density of inclusions [24].

SECM is a scanning electrochemical probe that detects amperometrically surface-generated electroactive ions or molecules in the solution phase as a function of spatial location with an electrochemically sensitive or ion-selective ultramicroelectrode tip. It has also been used to make high-resolution chemical concentration maps of corroding metal surfaces, and it was found that minor spatial fluctuations on the pA scale in passive current density of type 304 stainless steel in dilute aqueous chloride solution were related directly to the subsequent initiation of pitting corrosion [25]. SECM measurements suggest that sulfide inclusions should be the initiation sites, although a definite coincidence could not be established [25]. SECM was also applied to evaluate the heterogeneity of a passive film formed on a pure iron electrode in deaerated pH 8.4 berate solution [26]. A probe current image of SECM was measured with a tip-generation/substrate-collection (TG/SC) mode in deaerated pH 8.4 berate solution. The difference in the thickness of passive films formed on two iron plates at different potentials could be evaluated from the probe current image [26]. SECM was used to observe a natural and artificial distribution of electron transfer activity on glassy carbon electrodes [27]. A large (sevenfold) spread in rate constant is found for randomly sampled sites on polished, untreated glassy carbon surfaces. Direct-mode oxidation with the SECM tip was used to produce small regions of oxidized carbon on a polished surface. A large increase in the electron transfer rate for the Fe(II/III) ion is observed on the locally oxidized carbon surface compared to the unoxidized region [27]. SECM with glass-insulated platinum–lidium microelectrodes was used to measure the surface concentration of the chemical species generated by a specimen electrode as a function of the specimen's electrode potential [28]. The resulting surface concentration–potential curves contain information about the microscopically local electron-transfer kinetics of a reaction occurring at a kinetically heterogeneous electrode. The quality of the curves and the spatial resolution were measured as a function of the specimen electrode scan rate using ferrocyanide as the electroactive species [28]. SECM has also been utilized in biological systems. Scanning bioelectrochemical microscopy is shown to be extremely useful for the mapping of localized biological activity and the monitoring of dynamic biological events [29].

It should be noted that scanning probe techniques, including LEIS and SECM, commonly operate in a relatively specific and localized area, and thus the scan image may not necessarily represent the full details of an electrode process that involves different reactions occurring simultaneously over distinctively separate electrode areas. In the investigation of corrosion, it is difficult to ensure that a scanning tip is positioned correctly over a pit precursor unless the precursor is generated by the probe tip itself.

Various other surface analytical techniques have also been used in probing heterogeneous electrode–electrolyte interfaces. For example, a technique for microscopic imaging of electrochemically active surfaces has been introduced by combining concepts of probe microscopy and advances in mass spectrometry [30]. The technique is based on a miniature electrochemical flow cell scanner. A liquid feed stream containing a redox component is introduced in the vicinity of the location examined through the annulus of a coaxial capillary set. The incoming reagent interacts with the target location, and the product stream generated is transferred through the inner capillary to an electrospray mass spectrometer. Thus, a multicomponent potential-dependent image of the products' distribution versus the location on the electrode is generated [30].

An electrochemical microcell technique has enabled local electrochemical measurement such as local electrochemical impedance spectroscopy and cyclic voltammetry, providing a unique way of performing the investigation of heterogeneities on electrode surfaces [31]. Local electrochemical measurements on an artificially patinated copper surface have made it possible to distinguish areas of different reactivity even when an analysis of the surface revealed a homogeneous chemical composition of patina. Local measurements with an electrochemical microcell showed the presence of small defects on the copper patina layer [31].

Electrochemical in situ fluorescence microscopy was used to measure the surface heterogeneity of DNA self-assembled monolayers on gold electrodes [32]. Gold surfaces modified by single- or double-stranded DNA self-assembled monolayers were shown to produce heterogeneous surface packing densities through the use of electrochemical studies coupled with fluorescence imaging. The difference in fluorescence intensity measured from these regions was dramatic. More important, a regional variance in fluorescence intensity in response to electrochemical potential was observed, the large aggregates showing a significantly different modulation of fluorescence intensity than that of the monolayer-coated regions [32].

Electrostatic force microscopy was used to investigate the cell surface electrochemical heterogeneity of Fe(III)-reducing bacteria [3]. The pH dependence of electrostatic force microscopy image contrast paralleled the pattern of cell surface charge development inferred from titration experiments. However, quantitative analysis of high-contrast regions in the images yielded lower surface charge values than those anticipated from the titration data. A particularly high degree of electrochemical heterogeneity was found to exist within the cell wall and at the cell surface of the bacteria [3].

Arrays of platinum, gold, or glassy carbon microelectrodes are commonly used in electroanalytical applications because they provide circumstances in which a large and easily measured total current output is obtained while retaining many of the advantageous features possessed by the individual microelectrodes [33]. Although these microelectrode arrays are not designed for simulating or detecting electrochemical heterogeneity over practical macroelectrode surfaces, they are useful for studying electrode inhomogeneity in certain situations. For example, individually addressable carbon microelectrode arrays have been used for the spatial and temporal resolution of neurotransmitter release from single pheochromocytoma cells. Amperometric results show that subcellular heterogeneity in single-cell exocytosis can be detected electrochemically with these electrode arrays [34]. Carbon ring microelectrodes were utilized in neurochemistry for electrochemical monitoring of exocytotic events from single cells. Subcellular temporal heterogeneities in exocytosis (i.e., cold spots vs. hot spots) were detected successfully using these carbon ring microelectrodes [35].

Each technique has advantages and limitations; for this reason, different techniques are often combined and applied in a novel and synergistic manner. Wang et al. utilized various local electrochemical techniques, including the microelectrode, the scanning vibrating electrode, the wire beam electrode, scanning electrochemical microscopy, localized electrochemical impedance spectroscopy, scanning Kelvin probe force microscopy, and numerical simulation to study microelectrochemical heterogeneity of complex biofilm–metal interfaces. The advantages and limitations of these techniques have been summarized and discussed [36].

8.2 ELECTROCHEMICAL NOISE GENERATION FROM ELECTROCHEMICAL HETEROGENEITY

Random noise and its effect on physical systems has attracted interdisciplinary research interests since Einstein's work on the random fluctuations of tiny particles in a fluid (Brownian motion), and Johnson's study of the origin of noise in an electrical circuits. In an electrochemical system, spontaneous potential and current fluctuations (i.e., electrochemical noise) have received considerable attention since Iverson's study on electrode potential fluctuations and corrosion processes [37]. During the past three decades, two major applications of electrochemical noise

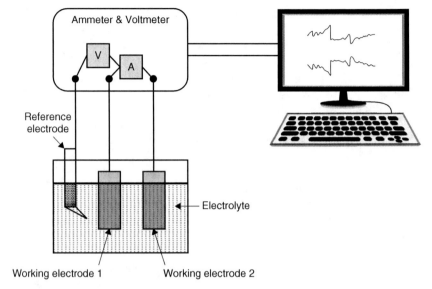

Figure 8.1 Experimental setup for detecting potential and current noise from two identical corroding electrodes.

analysis have been developed. The first is the noise resistance method, which was developed to determine general corrosion rates using noise resistance [38]. Another is the noise signatures method, which was proposed to detect localized corrosion by recognizing characteristic noise patterns (i.e., noise signatures) in the time domain [39] or the frequency domain [40]. The noise resistance method has been confirmed experimentally and theoretically [41–44], but the noise signature method, which has attracted the most interest due to it makes localized corrosion identification and quantification possible, remains a rather controversial issue. The noise signatures suggested as being indicators of localized corrosion often are not verified by other researchers. Figure 8.1 shows an experimental setup for detecting potential and current noise from two identical corroding electrodes.

Electrochemical heterogeneity has been recognized to play a critical role in electrode noise [45], and localized corrosion processes are believed to give rise to natural transient events [39,40]. The identification of noise signatures have been undertaken by a variety of means; for example, examination of the time record trace may give an indication of the types of processes occurring. The initiation of pitting is believed to be characterized by sharp fluctuations in potential and current. Typical potential fluctuation patterns of pitting initiation are illustrated in Figure 8.2.

Although progress has been made over the past decades on understanding electrochemical noise, there are still unanswered questions. First, the origin and generation mechanisms of electrochemical noise are not yet fully understood, although a number of noise-generating processes, such as metastable pitting, turbulent mass transport, particle impact, bubble nucleation, and separation, have been identified

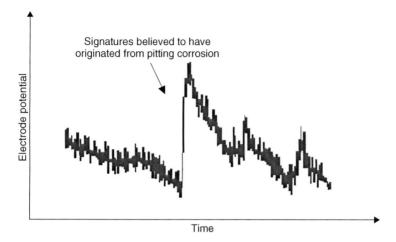

Figure 8.2   Typical potential noise from an electrode undergoing pitting corrosion. (From [46].)

as possible sources of electrode noise. Since there are difficulties in correlating noise signatures directly with localized corrosion activities occurring at a specific location on an electrode surface, many workers studying the application of noise signatures to the identification of localized corrosion have tried only to relate certain characteristic noise features they observed during certain periods in the experiment to localized corrosion, such as pits identified visually or microscopically after an experiment was complete. The expectation is that the number of peaks in potential fluctuation data for a certain immersion period could equal the number of pits observed after immersion using an optical microscope. However, it is well known that only a small portion of the peaks observable in potential fluctuation data lead to stable corrosion pits, and thus this approach is valid only if there are means of identifying the valid peaks that lead to the formation of stable pits. Furthermore, the sensitivity of conventional noise detection is often not high enough to recognize the relatively small noise activities associated with the initiation of tiny localized corrosion sites. This is because the traditional noise detection method using a one-piece electrode (or two short-circuited identical electrodes) measures only a mixed or averaged potential and its fluctuations over the entire electrode surface. The initiation of pitting corrosion usually involves only a small electrode area; therefore, such events could only result in a very small and often invisible fluctuation in an overall mixed or averaged electrode potential [47].

To overcome these difficulties, electrode arrays such as the WBE has been used to study localized corrosion-induced noise signatures [47]. An array of nickel electrodes undergoing electrodissolution in sulfuric acid was used in experiments on emerging coherence through weak global coupling of populations of electrochemical oscillators with heterogeneities [48]. Legat and Dolecek have reported two characteristic fluctuations of the currents measured from the microelectrode arrays and proposed that slower fluctuations were generated by general corrosion of the electrodes, whereas the shorter transients were very probably generated by the initiation of pits [49]. The WBE exhibits two special features that could be very valuable for such noise signature studies:

1. The addressable multielectrode structure allows a WBE to measure the local potential and its fluctuations. For example, if localized corrosion is initiated on a WBE surface due to local breakdown of the protective surface film, a sudden potential change could occur at locally broken areas. Although such an event may result in only small fluctuations in the overall electrode potential, it could result in significant local potential fluctuations that can be detected using a WBE. For this reason, the sensitivity of noise detection could be improved.

2. A WBE could enable the direct correlation of noise activities with a specific location on the WBE surface (i.e., to relate electrode noise to its origin).

An experiment was carried out using a stainless steel 316L WBE exposed to a corrosive environment containing $FeCl_3$ under open-circuit conditions (Figure 8.3) for detecting potential noise over a WBE surface and also for mapping galvanic currents flowing among coupled wires in the WBE [47]. Electrode potential noise was recorded by measuring the open-circuit potential of the WBE against a saturated calomel reference electrode using an automatic voltmeter. The same setup was used to measure galvanic current distribution maps over the WBE surface without interrupting the continuous electrode potential noise measurements. Figure 8.4 shows

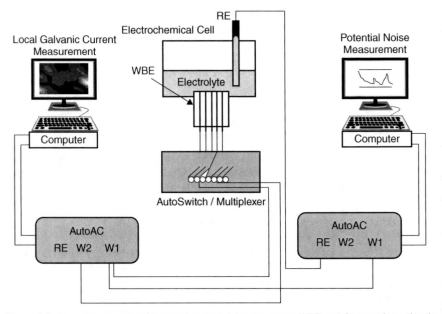

Figure 8.3 Experimental setup for detecting potential noise over a WBE and for mapping galvanic currents flowing into and out of each wire in the WBE.

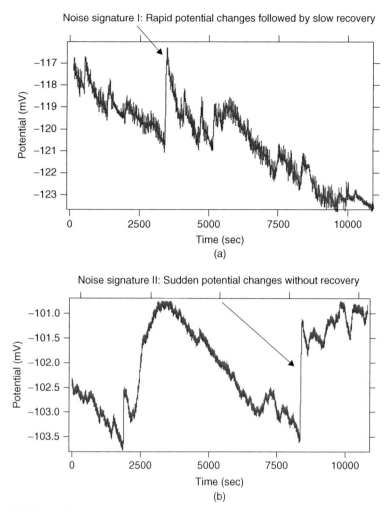

Figure 8.4   Potential noise measured from a stainless steel 316L WBE after exposure to 6% FeCl₃ solution for (a) 6 to 9 hours and (b) 87 to 90 hours.

typical potential noise measured from a stainless steel WBE after being exposed to a 6% $FeCl_3$ solution for various periods. Figure 8.5 shows some galvanic current distribution maps measured from the same WBE surface over relevant periods of exposure.

In this experiment, several characteristic potential noise signatures were identified [47]. Noise signature I in Figure 8.4a has a characteristic peak of rapid potential transient, toward a less negative direction, followed by exponential recovery. As shown in Figure 8.5, this noise signature correlated with the disappearance of a major anode over a 30-minute time window (between 4 hours 10 minutes and 4 hours 40 minutes) during period B. This result suggests that the origin of noise

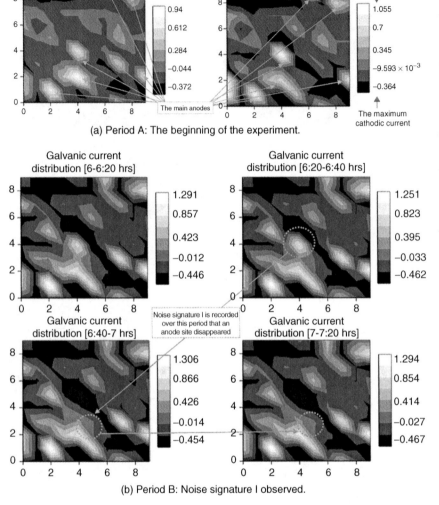

(a) Period A: The beginning of the experiment.

(b) Period B: Noise signature I observed.

Figure 8.5 Galvanic current (mA/cm²) distribution map obtained from an ss316L WBE after exposure to 6% FeCl₃ solution for various periods.

signature I in this particular experimental condition was not the conventionally believed passive film rupture process, but the anode disappearance process. The disappearance of a major anodic site led to a decrease in anodic area and thus caused a sudden shift of electrode potential to the less negative direction. Noise signature II in Figure 8.4b is featured with the characteristic pattern of rapid potential transient (also to the less negative direction) followed by partial or no recovery. As shown in Figure 8.5, this noise signature corresponded to a massive disappearance of anodic sites, leading to the formation of a few major stable anodes. During this period, the

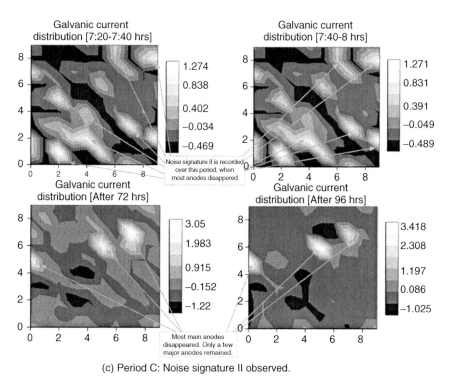

(c) Period C: Noise signature II observed.

Figure 8.5 *(Continued)*

anodic dissolution current of the stable anode increased significantly, from about 3.05 mA/cm^2 to over 10 mA/cm^2. This experiment demonstrated clearly that noise signatures I and II are indicators of minor anode disappearance, leading to accelerated dissolution of major anodes. Obviously, recognition of such noise signatures could be employed as a means of early detection and prediction of pitting corrosion. This result shows an interesting mechanism of pitting corrosion initiation and propagation that is different from conventional passive film breakdown–based pitting mechanisms [47].

8.3 HARVESTING ELECTRICAL POWER FROM ELECTROCHEMICAL HETEROGENEITY USING A WBE

One application of electrochemistry heterogeneity is power generation. Various forms of energy generation and storage devices, such as batteries, fuel cells, and solar cells, have been developed to make use of various forms of significant electrochemical heterogeneity. Localized corrosion is a significant heterogeneous electrode process and thus could be used as a natural power source if a suitable device for harvesting electrical power from localized corrosion processes is available. As an

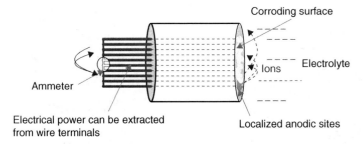

Corroding surface

Electrolyte

Ions

Ammeter

Electrical power can be extracted
from wire terminals

Localized anodic sites

Figure 8.6 Conceptual design of a method for extracting electrical power from localized corrosion.

example, the WBE is employed as a novel means of generating and extracting electrical power out of a corroding metal surface where electrochemical heterogeneity is initiated and propagated. Figure 8.6 illustrates a conceptual design for electrical power extraction from localized corrosion using a WBE device.

Corrosion, especially localized forms of corrosion, is rarely reported to be a beneficial phenomenon, although it has been used in research to produce porous materials. A galvanic cell is often used to describe the fundamental electrochemical mechanism of aqueous corrosion. However, unlike a conventional galvanic cell such as a battery or a fuel cell, where spontaneous electrochemical reactions occur at a separated anode and cathode, aqueous corrosion usually occurs over a single piece of metal surface where anodic and cathodic reactions occur concurrently. It is well known that uniform corrosion is the result of the operation of many tiny galvanic cells that have numerous microscopic anodes and cathodes distributed randomly over a metal surface. The total anodic current equals the total cathodic current, and the net measurable current is zero. The anode and cathode locations change dynamically, and thus any given area on the metal surface would act as both an anode and a cathode over any extended period of time, resulting in uniform dissolution of the metal surface. In uniform corrosion, electrical power carried by electrons is wasted during the transfer of electrons from numerous tiny anodes to cathodes. Obviously, the extraction of energy from uniform corrosion is difficult, if not impossible.

Localized corrosion such as pitting has a different mechanism, in which localized dissolution of metals occurs due to the formation of stable anodic sites. The root cause of localized corrosion is electrochemical heterogeneity. Significant electrochemical heterogeneity would naturally initiate and propagate over a metal surface when it is exposed to certain corrosive media. Localized corrosion processes generate electric currents flowing between macroscopic-scale anodic-to-cathodic areas. Unfortunately, electrical energy carried by these currents is also wasted when electrons move through an electrode body. Obviously, the extraction of energy from localized corrosion cells is not possible if a conventional single-piece electrode is used.

Figure 8.6 shows the design concept of a method that overcomes difficulties in extracting power from localized corrosion processes. The WBE is an array of minielectrodes (wires) that are insulated from each other by a thin insulating layer. The working surface of the WBE is integrated electrochemically by coupling all the terminals of the metal wires and by closely packing all the wires at the solid–electrolyte interface. Figure 8.7 shows a power generation and extraction experiment using a steel WBE. The WBE surface is exposed to an Evans solution, a well-known environment in which electrochemical heterogeneity is expected to be initiated and propagated, leading to localized corrosion. The electrically coupled and close-packed array in the WBE enables effective interactions between local anodes and cathodes and ensures low resistance to ions and electron movements. Electric currents flowing through wire terminals can be measured and utilized as electrical power. This can be done by identifying and tying all anodic wire terminals together and then connecting them to cathodic wire terminals via an ammeter and/or an electrical load. The WBE used in this work was made from 100 identical mild steel (UNS No. G10350) wires embedded in epoxy resin and insulated from each other by a very thin epoxy layer. Each wire had a diameter of 0.18 cm and acted as both a minielectrode and a corrosion substrate. The working area (the area occupied by the wire array) was approximately 3.24 cm^2 (1.8 cm × 1.8 cm). The total metallic area was approximately 2.55 cm^2. The working surface of the WBE was abraded with 400–, 800–, and 1200-grit silicon carbide paper and cleaned with ethanol and isopropanol. The working surface was totally immersed in Evans solution containing 0.017 M sodium chloride and 0.008 M sodium carbonate under static conditions at 20°C to allow corrosion to occur. During exposure, all the wire terminals of a WBE were connected together so that electrons could move freely between wires, similar to the movement in a single piece of mild steel electrode.

Corrosion currents flowing between anodic and cathodic wires were measured and monitored over an extended period of time in order to understand power generation behavior with the progress of localized corrosion processes. The measurement

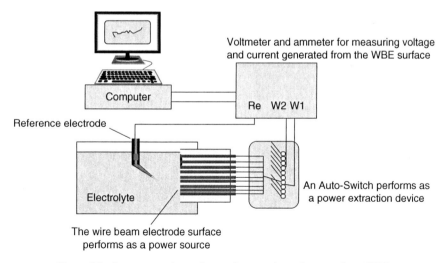

Figure 8.7   Power generation and extraction experimental setup using a WBE.

of current distribution was carried out by connecting an automatic zero-resistance ammeter (AutoZRA, ACM Instruments, England) between one wire terminal and all the other terminals shorted together using a computer-controlled automatic switch device (custom made) to measure galvanic currents flowing between each wire and the wire beam system. The corrosion potential distribution was obtained by measuring the open-circuit potential of each wire of a WBE against an Ag/AgCl reference electrode. The terminals of wires in the electrode were temporarily disconnected from the WBE system and connected in sequence to a digital voltmeter using the automatic switch device. Experimentally measured raw current and potential data were transferred to analysis software written under a Mathcad environment (Mathcad Professional 7, MathSoft, Inc., Massachusetts). Maps showing the corrosion potential distribution and galvanic current distribution were plotted. Readers are referred to the literature for technical and analysis details [50,51].

Electrode potential data are useful for an understanding of the thermodynamics and possible voltage output of a particular corrosion system, while current values are useful for determining the kinetics and expected current output of the system. Figures 8.8 and 8.9 show corrosion potential and galvanic current distribution maps measured from a mild steel WBE surface immersed in Evans solution. It is shown clearly in Figure 8.8 that the WBE surface was thermodynamically heterogeneous, as indicated by the nonuniform corrosion potential distribution, and this electrochemical heterogeneity was propagated with the extension of electrode immersion. The current maps in Figure 8.9 correlated well with the potential maps in Figure 8.8.

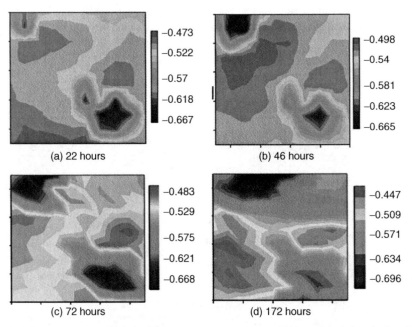

(a) 22 hours   (b) 46 hours

(c) 72 hours   (d) 172 hours

Figure 8.8 Corrosion potential distribution maps measured over a WBE surface during the first 172 hours' exposure (potential in volts). (*See insert for color representation of the figure.*)

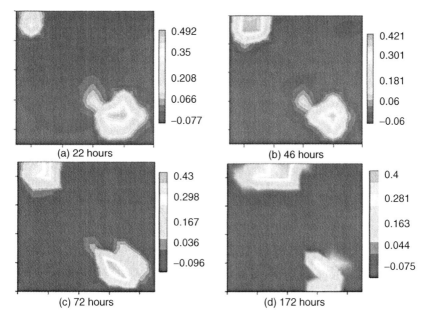

Figure 8.9 Corrosion current distribution maps measured over a WBE surface during the first 172 hours' exposure (in mA/cm^2). (*See insert for color representation of the figure.*)

With the extension of exposure (Figures 8.8 and 8.9), electrochemical heterogeneity propagated on the WBE surface, forming several major anodic and cathodic zones. The thermodynamic differences over the electrode surface (indicated by potential differences) resulted in kinetic heterogeneity (indicated by current differences). As a result, localized corrosion initiated, propagated, and became highly concentrated on a few anodic sites. The vast majority of the WBE surface areas behaved as cathodes. The propagation of electrochemical heterogeneity and localized corrosion appeared to be a self-catalytic process, a phenomenon that had been observed previously in localized corrosion such as pitting corrosion [52].

In this localized corrosion process, electrons transferred continuously from the anodic to the cathodic areas through the electrode terminals. This current can be extracted by selecting and separating anodic and cathodic wires. Figure 8.10 shows total current generated from this localized corrosion power source. The total current is the sum of all anodic or cathodic currents flowing in the wire terminals. It is determined by adding all anodic or cathodic currents measured at any point of the experiment since the sum of all anodic currents equals the sum of all cathodic currents. It is shown that the total current output from this localized corrosion system increased with the extension of localized corrosion and reached around 0.084 mA after 172 hours. As an accumulated result, the electrode surface was locally corroded, with the appearance of waterline and pitting corrosion.

Figure 8.11 shows open-circuit voltage output over the period of a localized corrosion experiment. It is shown that the open-circuit voltage output from this

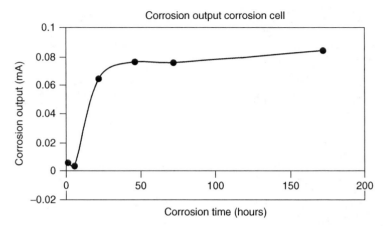

Figure 8.10 Total current output from a WBE localized corrosion system over the period of localized corrosion.

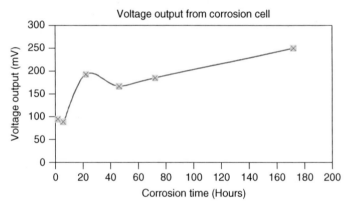

Figure 8.11 Open-circuit voltage output from a WBE localized corrosion system over the period of localized corrosion.

localized corrosion system increased with the extension of localized corrosion and reached around 250 mV after 172 hours. It is expected that the WBE surface would reach a steady state and that miniequilibria would be established over various locations on the WBE surface.

Obviously, the capability of electrical power generation from this WBE localized corrosion system is relatively small compared to that of conventional batteries and fuel cells. However, the capability of this battery could be improved if a different electrode material, such as magnesium, were used. This new power source could have advantages in certain applications. For example, it could be used to power implants such as cardiac pacemakers, sensors, and miniature drug pumps in the human body. This WBE localized corrosion power source could use human blood, body fluid, and sweat as an electrolyte. It could be developed into a simple and long-lasting form of power source.

8.4   FURTHER RESEARCH ISSUES IN ELECTROCHEMICAL HETEROGENEITY

The fundamental significance of electrode inhomogeneity and electrochemical heterogeneity may need to be more widely recognized as a key factor affecting a multidisciplinary field dealing with heterogeneous electrode–solution interfaces such as electrocatalysis, fuel cells, solar cells, alloy plating, nanofabrication, biology, bioelectrochemistry, biosensors, medical engineering, interaction of tissue with metal implants. Electrochemical heterogeneity could be a satisfactory explanation for nonideal electrode behavior and could be used for the effective control of electrode processes.

It should be recongized that electrochemical heterogeneity could occur at various spatial and time scales, and thus different techniques could be used to probe different heterogeneous electrode processes. For example, scanning probe techniques such as scanning Kelvin probe force microscopy and scanning electrochemical microscopy could be useful for probe nanoscale electrochemical heterogeneity; the scanning Kelvin probe, the scanning vibrating electrode technique, local electrochemical impedance spectroscopy, and microelectrode arrays could be used for probing microscale electrochemical heterogeneity, while the wire beam electrode could be more suitable for probing macroscale heterogeneous electrode processes. Since the scale of electrochemical heterogeneity could change with the propagation of heterogeneous electrode processes, these techniques could be employed in various combinations.

The WBE has been used successfully to measure relatively slow heterogeneous electrode processes such as the growth of localized corrosion. Its instrumentation in its present form has major limitations in measuring rapid electrode processes since it takes approximately 15 minutes to produce a potential or current distribution map of the type shown in Figures 3.6 and 3.7. Owing to the limitations in data accumulation speed, the method has been applied only to static or quasi-equilibrated systems. Considering the fact that most critical heterogeneous electrochemical processes, such as the initiation of localized corrosion, the breakdown of surface film, and the origination of electrochemical noise, involve dynamic localized electrochemical activities, this limitation would seriously affect the applicability of the method in studies that require high spatial and time resolution. Instrumentation that is able to produce a current flow video on a WBE surface would be highly desirable. The method could be used in conjunction with scanning probes such as atomic force microscopy and scanning electrochemical microscopy to enable electrochemists to view real-time video of dynamically evolving electrochemical heterogeneity simultaneously with images of localized surface chemistry and morphology.

Many unsolved fundamental electrochemistry issues, such as those identified by Gileadi [53], could also be related to electrochemical heterogeneity at the nano- or microscales. For example, on an electrode surface with heterogeneous adsorption, the electrostatic field at the adsorption site could be highly distorted, leading to nanoscale electrochemical heterogeneity. In this case, the overpotential imposed might not apply at the point where the reaction takes place, and there could be uncertainty in the symmetry factor values in the Tafel equation in different areas of the Helmholtz plane. This could be responsible for the fact that truly linear Tafel lines are difficult to observe unless a hanging or dropping mercury electrode is used for careful determination of the current–potential relationship [53]. At the macroscopic scale, for example, the positioning of the tip of a Luggin capillary close to the working electrode could lead to a capillary that screens the electrode and distorts the uniformity of the current distribution. This could lead to grossly distorted electrochemical measurement results [53].

REFERENCES

1. N. Campbell and J. Reece, *Biology*, 7th ed., Pearson Benjamin Cummings, San Francises, CA, 2005.

2. O. Sterkers, G. Saumon, P. Tran Ba Huy, E. Ferrary, and C. Amiel, Electrochemical heterogeneity of the cochlear endolymph: effect of acetazolamide, *Renal Physiology*, 246 (1984), F47–F53.

3. I. Sokolov, D. S. Smith, G. S. Henderson, Y. A. Gorby, and F. G. Ferris, Cell surface electrochemical heterogeneity of the Fe(III)-reducing bacteria *Shewanella putrefaciens*, *Environmental Science and Technology*, 35 (2001), 341–347.

4. B. D. VanVeen, W. vanDrongelen, M. Yuchtman, and A. Suzuki, Localization of brain electrical activity via linearly constrained minimum variance spatial filtering, *IEEE Transactions on Biomedical Engineering*, 44 (1997), 867–880.

5. P. Yuan and S. F. Liu, Numerical analysis of temperature and current density distribution of a planar solid oxide fuel cell unit with nonuniform inlet flow, *Numerical Heat Transfer A*, 51 (2007), 941–957.

6. M. M. Mench and C. Y. H. Wang, An in situ method for determination of current distribution in PEM fuel cells applied to a direct methanol fuel cell, *Journal of the Electrochemical Society*, 150 (2003), A79–A85.

7. J. J. Hwang, W. R. Chang, R. G. Peng, P. Y. Chen, and A. Su, Experimental and numerical studies of local current mapping on a PEM fuel cell, *International Journal of Hydrogen Energy*, 33 (2008), 5718–5727.

8. A. Saatchi and A. Aghajani, Interference problems and nonuniform potentials in cathodic protection of a complex installation, *Materials Performance*, 44 (2005), 22–25.

9. X. Chen, X. G. Li, C. W. Du, and Y. F. Cheng, Effect of cathodic protection on corrosion of pipeline steel under disbonded coating, *Corrosion Science*, 51 (2009), 2242–2245.

10. D. T. Chin, and G. M. Sabde, Current distribution and electrochemical environment in a cathodically protected crevice, *Corrosion*, 55 (1999), 229–237.

11. H. M. Yang, D. H. Kim, D. Zhu, and K. Wang, Improvement of deposition uniformity in alloy electroforming for revolving parts, *International Journal of Machine Tools and Manufacture*, 48 (2007), 329–337.

12. S. John, V. Ananth, and T. Vasudevan, Improving the deposit distribution during electroforming of complicated shapes, *Bulletin of Electrochemistry*, 15 (1999), 202–204.

13. G. S. Nagarajan, C. H. Ho, and J. W. VanZee, A mathematical model for predicting nonuniform electrochemical impregnation of nickel hydroxide, *Journal of the Electrochemical Society*, 144 (1997), 503–510.

14. A. I. Dikusar, O. O. Redkozubova, S. P. Yushchenko, L. B. Kriksunov, and D. Harris, Role of hydrodynamic conditions in the distribution of anodic dissolution rates in cavity etching regions during electrochemical micromachining of partially insulated surfaces, *Russian Journal of Electrochemistry*, 39 (2003), 1073–1077.

15. V. M. Kushnarenko, B. V. Perunov, V. F. Yakovenko, and I. V. Yureskul, Electrochemical heterogeneity of pipe steel weld joints in hydrogen-sulfide-containing media, *Materials Science*, 21 (1985), 279–281.

16. K. Ray and R. L. McCreery, Spatially resolved Raman spectroscopy of carbon electrode surfaces: observations of structural and chemical heterogeneity, *Analytical Chemistry*, 69 (1997), 4680–4687.

17. J. S. Gnanaraj, M. D. Levi, E. Levi, G. Salitra, D. Aurbach, J. E. Fischer, and A. Claye, Comparison between the electrochemical behavior of disordered carbons and graphite electrodes in connection with their structure, *Journal of the Electrochemical Society*, 148 (2001), A525–A536.

18. Y. J. Tan, Sensing electrode inhomogeneity and electrochemical heterogeneity using an electrochemically integrated multi-electrode array, *Journal of the Electrochemical Society*, 156 (2009), C195–C208.

19. Y. J. Tan, Sensing pitting corrosion by means of electrochemical noise detection and analysis, *Sensors and Actuators B*, 139 (2009), 688–698.

20. P. Cordoba-Torres, R. P. Nogueira, L. de Miranda, L. Brenig, J. Wallenborn, and V. Fairen, Cellular automaton simulation of a simple corrosion mechanism: mesoscopic heterogeneity versus macroscopic homogeneity, *Electrochimica Acta*, 46 (2001), 2975–2989.

21. M. W. Wittmann, R. B. Leggat, and S. R. Taylor, The detection and mapping of defects in organic coatings using local electrochemical impedance methods, *Journal of the Electrochemical Society*, 146 (1999), 4071–4075.

22. F. Zou and D. Thierry, Localized electrochemical impedance spectroscopy for studying the degradation of organic coatings, *Electrochimica Acta*, 42 (1997), 3293–3301.

23. V. M Huang, S. L. Wu, M. E. Orazem, N. Pebere, B. Tribollet, and V. Vivier, Local electrochemical impedance spectroscopy: a review and some recent developments, *Electrochimica Acta*, 56 (2011), 8048–8057.

24. H. Krawiec, V. Vignal, and J. Banas, Local electrochemical impedance measurements on inclusion-containing stainless steels using microcapillary-based techniques, *Electrochimica Acta*, 54 (2009), 6070–6074.

25. Y. Y. Zhu, and D. E. Williams, Scanning electrochemical microscopic observation of a precursor state to pitting corrosion of stainless steel, *Journal of the Electrochemical Society*, 144 (1997) L43–L45.

26. K. Fushimi, K. Azumi, and M. Seo, Evaluation of heterogeneity in thickness of passive films on pure iron by scanning electrochemical microscopy, *ISIJ International*, 39 (1999), 346–351.

27. R. C. Tenent, and D. O. Wipf, Local electron transfer rate measurements on modified and unmodified glassy carbon electrodes, *Journal of Solid State Electrochemistry*, 13 (2009), 583–590.

28. R. C. Engstrom, B. Small, and L. Kattan, Observation of microscopically local electron-transfer kinetics with scanning electrochemical microscopy, *Analytical Chemistry*, 64 (1992), 241–244.

29. J. Wang, Scanning probe microscopies for high-resolution characterization of electrochemical sensors, *Analyst*, 117 (1992), 1231–1233.

30. A. D. Modestov, S. Srebnik, O. Lev, and J. Gun, Scanning capillary microscopy/mass spectrometry for mapping spatial electrochemical activity of electrodes, *Analytical Chemistry*, 73 (2001), 4229–4240.

31. M. M. Mennucci, M. Sanchez-Moreno, I. V. Aoki, M.-C. Bernard, H. G. de Melo, S. Joiret, and V. Vivier, Local electrochemical investigation of copper patina, *Journal of Solid State Electrochemistry*, 16 (2012), 109–116.

32. J. N. Murphy, A. K. H. Cheng, H. Z. Yu, and D. Bizzotto, On the nature of DNA self-assembled monolayers on Au: measuring surface heterogeneity with electrochemical in situ fluorescence microscopy, *Journal of the Americal Chemical Society*, 131 (2009), 4042–4050.

33. K. Aoki, Theory of microelectrodes, *Electroanalysis*, 5 (1993), 627–639.

34. B. Zhang, K. L. Adams, S. J. Luber, D. J. Eves, M. L. Heien, and A. G. Ewing, Spatially and temporally resolved single-cell exocytosis utilizing individually addressable carbon microelectrode arrays, *Analytical Chemistry*, 80 (2008), 1394–1400.

35. Y. Q. Lin, R. Trouillon, M. I. Svensson, J. D. Keighron, A. S. Cans, and A. G. Ewing, Carbon-ring microelectrode arrays for electrochemical imaging of single cell exocytosis: fabrication and characterization, *Analytical Chemistry*, 84 (2012), 2949–2954.

36. W. Wang, X. Zhang, and J. Wan, Heterogeneous electrochemical characteristics of biofilm/metal interface and local electrochemical techniques used for this purpose, *Materials and Corrosion*, 60 (2009), 957–962.

37. W. P. Iverson, Transient voltage changes produced in corroding metals and alloys, *Journal of the Electrochemical Society*, Electrochemical Science (1968), 617–618.

38. D. A. Eden, K. Hladky, D. G. John, and J. L. Dawson, Electrochemical noise resistance, Paper 274, presented at Corrosion '86, NACE, Houston, TX, 1986.

39. K. Hladky and J. L. Dawson, The measurement of pitting corrosion using electrochemical noise, *Corrosion Science*, 21 (1981), 317–322.

40. K. Hladky and J. L. Dawson, The measurement of corrosion using electrochemical $1/f$ noise, *Corrosion Science*, 22 (1982), 231–237.

41. F. Mansfeld and H. Xiao, Electrochemical noise analysis of iron exposed to NaCl solutions of different corrosivity, *Journal of the Electrochemical Society*, 140 (1993), 2205.

42. Y. J. Tan, S. Bailey, and B. Kinsella, The monitoring of the formation and destruction of corrosion inhibitor films using electrochemical noise analysis, *Corrosion Science*, 38 (1996), 1681.

43. U. Bertocci, C. Gobrielli, F. Huet, and M. Keddam, Noise resistance applied to corrosion measurements: I. Theoretical analysis, *Journal of the Electrochemical Society*, 144 (1997), 31–37.

44. R. A. Cottis, Interpretation of electrochemical noise data, *Corrosion*, 57 (2001), 265–285.

45. J. Wojtowicz, in *Modern Aspects of Electrochemistry*, J. O'M. Bockris and B. E. Conway, Eds., No. 8, Butterworth, London, 1973, p. 52.

46. Y. J. Tan, S. Bailey, B. Kinsella, and A. Lowe, Analysis of electrochemical noise using Fourier transform, maximum entropy and wavelet methods, *Proceedings of Corrosion and Prevention '98*, Australasian Corrosion Association, Perth, WA, Australia, 1998.

47. Y. J. Tan, N. N. Aung, and T. Liu, Novel corrosion experiments using the wire beam electrode: I. Studying electrochemical noise signatures from localised corrosion processes, *Corrosion Science*, 48 (2006), 23–38.

48. Y. M. Zhai, I. Z. Kiss, and J. L. Hudson, Emerging coherence of oscillating chemical reactions on arrays: experiments and simulations, *Industrial and Engineering Chemistry Research*, 43 (2004), 315–326.

49. A. Legat and V. Dolecek, Corrosion monitoring system based on measurement of electrochemical noise, *Corrosion*, 51 (1995) 295–300.

50. Y. J. Tan, Corrosion science: a retrospective and current status, presented at the Electrochemical Society 201st Meeting, Philadelphia, G. S. Frankel, H. S. Isaacs, J. R. Scully, and J. D. Sinclair, Eds., 2002, PV2002-13.

51. Y. J. Tan, Monitoring localised corrosion processes and estimating localised corrosion rates by means of a wire beam electrode, *Corrosion*, 54 (1998) 403–413.

52. G. Fontana, *Corrosion Engineering*, 3rd ed., McGraw-Hill, New York, 1987.

53. E. Gileadi, Problems in interfacial electrochemistry that have been swept under the carpet, *Journal of Solid State Electrochemistry*, 15 (2011), 1359–1371.

Index

Heterogeneous Electrode Processes and Localized Corrosion, First Edition. Yongjun Tan.
© 2013 John Wiley & Sons, Inc. Published 2013 by John Wiley & Sons, Inc.